W9-BMF-907

Climate Confusion

CLIMATE CONFUSION

How Global Warming Hysteria Leads to Bad Science, Pandering Politicians, and Misguided Policies That Hurt the Poor

BY

ROY W. SPENCER

ENCOUNTER BOOKS

New York · London

First edition published in 2008 by Encounter Books, an activity of Encounter for Culture and Education, Inc., a nonprofit, tax exempt corporation.

Encounter Books website address: www.encounterbooks.com

Manufactured in the United States and printed on acid-free paper. The paper used in this publication meets the minimum requirements of ANSI/NISO Z39.48_1992 (R 1997) (*Permanence of Paper*).

FIRST EDITION

LIBRARY OF CONGRESS CATALOGING-IN-PUBLICATION DATA

Spencer, Roy W.
Climate confusion : how global warming hysteria leads to bad science, pandering politicians, and misguided policies that hurt the poor / by Roy W. Spencer.
p. cm.
Includes index.
ISBN-13: 978-1-59403-210-3 (hardcover : alk. paper)
ISBN-10: 1-59403-210-6 (hardcover : alk. paper)
1. Global warming. I. Title.
QC981.8.G56S645 2008
363.738'74-dc22
2008001915

10 9 8 7 6 5 4

Contents

Preface VII

Prologue 1

Chapter 1: Global Warming Hysteria 11
All natural disasters are now caused by global warming.

Chapter 2: Science Isn't Truth 35
What we know isn't necessarily so.

Chapter 3: How Weather Works 45
The Mission: to move heat from where there is more to where there is less.

Chapter 4: How Global Warming (Allegedly) Works 62
The popular explanation, and why it is probably wrong.

Chapter 5: The Scientists' Faith, the Environmentalists' Religion 85
Belief in dangerous global warming is more faith than science.

Chapter 6: It's Economics, Stupid 103
Views on what should be done about global warming are usually related to what we believe about economics and wealth.

Chapter 7: The Politics of Climate Change 124
No other public issue has so much potential for abuse of authority.

Chapter 8: Dumb Global Warming Solutions 139
Are they really serious about fixing global warming, or are they just pulling our leg?

Chapter 9: Less Dumb Global Warming Solutions 160
New energy technologies of the future are the only hope to "save" us from the threats posed by global warming.

Chapter 10: Summary 171

Epilogue 183

Illustration credits 185

Index 187

Preface

Manmade global warming is a danger to humanity and the environment, and it must be stopped. This claim represents a leap of faith from what science tells us is theoretically possible, to a belief in worst-case scenarios in which Mother Earth punishes us for our sins against her.

Nowhere is this leap of faith better illustrated than in Al Gore's movie, *An Inconvenient Truth*. Dramatic video of weather events that occur naturally every day suddenly becomes evidence for global warming. Floods? Global warming. Droughts? Global Warming. Ice calving off of glaciers and falling into the ocean? Global warming. Hurricanes? Global warming. Do you see a pattern here? Global warming.

Not to be outdone, the actor Leonardo DiCaprio then hosted a ninety-minute documentary called *The 11th Hour*, in which nature has more human rights than humans do. I do so admire actors and actresses for their concern over our well-being.

Mr. Gore has admitted that the issue is a spiritual one for him. He is spreading the word, calling on humanity to avert the climatic cataclysm that is just around the corner. He is even training hundreds of disciples to go throughout the world, warning of the coming environmental apocalypse. All we need to do to be saved from the heat of global warming hell is to use more compact fluorescent light bulbs, hybrid cars, and purchase carbon offsets from his company.

Some schools and university professors have even required their students to watch Mr. Gore's movie. Our children are having nightmares because of what well-intentioned teachers are telling them about climate change. Fear is gradually replacing reason as a motivating force for societal change, and in today's world we know only too well what religious zealotry can do to a society.

These "global warnings" are accompanied by an increasing tone of urgency. Not long ago we were told humanity had fifty years to solve the global warming problem. Then, we heard we have only ten years to change our polluting ways. Now, some are claiming we have only five years left. Soon, we'll be talking about sending a Terminator back through time to fix the problem for us. Maybe the Governor of California can help us with that.

As an atmospheric scientist, I will admit that harmful global warming is indeed a possible outcome of mankind's emissions of greenhouse gases from fossil fuel use. But extraordinary claims require extraordinary evidence. It is relatively easy to construct a simple computerized climate model that produces catastrophic global warming when its "greenhouse effect" is cranked up a notch with more carbon dioxide. I've even done it in an Excel spreadsheet. But it is much more difficult to get even a state-of-the-art climate model to behave realistically when it is compared to the real climate system.

It would be natural for a climate scientist like me to write a book about the science of global warming alone. But my motivation for writing this book is not only because I think the public is getting a biased message on the science. It's also because the human cost of what we might do as a policy response to global warming will be so high. Climate change is indeed a moral issue, but not the one that Mr. Gore claims it is.

Fortunately, amidst all of the frightful climate change news stories, those who are riding the global warming bandwagon are providing no small amount of entertainment. Movie stars, politicians-turned-movie stars, and famous musicians apparently don't notice the hypocrisy of calling for humanity to use less toilet paper while they fly their private jets from city to city. And what could be more ironic than the early 2007 trek to the North

Pole to raise awareness of global warming that had to be called off because of cold weather?

If you are interested in the global warming debate, but don't want another boring list of dry scientific facts, then you've come to the right place. The only thing that depresses me more than the thought of reading such a book is the thought of having to write one. While this book will indeed help you understand the reasons why global warming is unlikely to be a serious threat, it will also entertain you.

I believe that we need a large measure of humor when facing those who would kill us with their good intentions. I don't know about you, but I need a daily dose of humor just to keep my sanity. As that famous philosopher-turned-comedian-turned-actor Steve Martin used to say, "A day without laughter is like a day without sunshine, and a day without sunshine is like . . . night."

This is Doctor Bagshaw, discoverer of the infinitely expanding research grant.

Prologue

MARK TWAIN OBSERVED that "everyone talks about the weather, but no one does anything about it." Well, today's popular view is that we finally are doing something about the weather. We are making it worse.

Because of humanity's emissions of carbon dioxide from the burning of fossil fuels, many scientists are predicting dramatic weather changes ahead. Depending upon which scientists you believe, the extra carbon dioxide we are putting in the atmosphere could melt the Greenland and Antarctic ice sheets, flooding coastal locations worldwide. It could shut down the Atlantic Gulf Stream and oceanic thermohaline circulation, triggering the rapid onset of a new Ice Age. Global weather circulation changes could cause more severe floods and droughts, altering or even destroying entire ecosystems.

The fear of global warming has galvanized the environmental movement and has led to billions of dollars in federal expenditures to observe and understand the climate system. It has spun off popular movies and helped to solidify political movements, such as the Green Party in Germany. Even a former U.S. Vice President, Al Gore, has written books and made a movie addressing the problem. Global warming has given new purpose to the lives of entertainers and movie stars, some of whom have taken a special interest in the issue.

Oh, and we scientists who make our living off it think it's a pretty cool gig, too.

But now, the western world's fear of global warming and its effects has reached the point of being an obsession. The media is more than willing to spread, and even amplify, the fear that humanity is filling up the Earth, pushing it beyond its ability to sustain us. Mother Nature is suffering as a result of our sins, and humans are now being increasingly blamed for every hurricane, tornado, tsunami, earthquake, flood, and drought that occurs.

Art Bell's popular book *The Coming Global Superstorm* and its movie spin-off, *The Day After Tomorrow*, are good examples of the public's fascination with fears of global climate catastrophes. I would say that the coming global superstorm has already arrived —but it is a storm of hype and hysteria.

I believe that the environmental fears that have consumed the western world stem from two central beliefs. The first is that the Earth is fragile and needs to be protected, even to the detriment of humans if necessary. Many people feel like the climate system is being pushed beyond its limits, past some imaginary tipping point from which there will be no return.

The second belief is that the increasing wealth of nations is bad for the environment. Since technology and our desire for more stuff are to blame for environmental problems, we should renounce our modern lifestyle.

I will argue for exactly the opposite viewpoint: that the Earth is pretty resilient; and that only through mankind's ingenuity and freedom to create wealth do we solve, or at least minimize, environmental problems as they arise.

We have had no shortage of pessimistic environmental predictions over the last forty years. The birth of the modern environmental movement is usually traced to the publication of Rachel Carson's *Silent Spring*. A biologist, Carson was passionate about the dangers of the insecticide DDT, which was in widespread use at the time. One concern was that DDT was causing a thinning of egg shells in some birds; another was that DDT was causing problems throughout the food chain.

While Carson is still admired for paving the way for future generations of environmentalists, governmental policies resulting from her work have caused the deaths of literally millions of people by allowing malaria to thrive in Africa. Instead of greatly reducing the amount of DDT that was so indiscriminately sprayed on crops, governments banned the use of the pesticide altogether. That the most famous policy reaction to environmental concerns has caused so much human suffering should, by itself, make us wary of any sweeping efforts to "protect the environment."

While Carson's research dealt with the dangers of one particular insecticide, it wasn't long before predictions of more widespread doom from other human pressures on the environment began to appear. In Paul Ehrlich's 1968 book, *The Population Bomb*, Ehrlich predicted that worldwide crises in food supply and natural resource availability would occur by 1990. Huge famines and economic system failures were predicted, destabilizing social and political order in the world. The basic premise of the book was that, while available resources were growing linearly with time, the population of the Earth was growing faster, at a geometric rate. Eventually, the population pressure would be too much—"unsustainable" in today's environmentally-friendly lexicon.

The only problem with Ehrlich's premise was that it was not true, and the crises never materialized. This led to the economist Julian Simon winning a famous bet with Ehrlich over whether several natural resources would become less or more available between 1980 and 1990. Simon allowed Ehrlich to choose five metals that Ehrlich thought would go up in price. Ehrlich chose copper, chrome, nickel, tin, and tungsten. A decrease in resource availability would be measured as an increase in price. Ten years later, in 1990, Dr. Ehrlich was forced to write a check to Dr. Simon, since the cost of all of the metals had decreased over the previous ten years.

While Ehrlich was correct that the amount of raw material in the ground does go down as mankind removes it, Julian Simon noted that mankind always adapts. We become more efficient in our use of those materials, or we find replacement materials. Someday we might even be mining our landfills to recover and recycle discarded materials.

In fact, almost all known reserves of resources have actually grown faster than the population over time. Even the United Nations, which never saw a crisis it wouldn't take money for to fail at solving, has projected that the global population will level off in this century. But this hasn't prevented a variety of experts to continue to claim that humanity's current rate of consumption cannot be sustained.

Not every environmentalist has bought into predictions of global doom, though. In the late 1990s, a professor of statistics and self-proclaimed environmentalist decided to examine many of the environmentalists' claims. Bjorn Lomborg and his statistics students started investigating the data that environmentalists were basing their gloomy predictions of environmental disaster on. He thus embarked on his road to conversion from environmental worry-wart to an optimistic defender of capitalism and the future of both humanity and the Earth.

By almost every measure, Lomborg found that the state of humanity and the Earth has gradually improved, most noticeably in the last hundred years. On average, people are living longer, healthier, better-fed, and more prosperous lives than ever before. Many diseases have been eradicated, and the gradual spread of free markets around the world has led to more efficient and cleaner use of natural resources.

In his book *The Skeptical Environmentalist*, Lomborg makes it clear that there is still room for improvement in many areas. But the idea that "things are getting worse" is just plain wrong.

Even overpopulation is now much less of a concern than it used to be. As the developing countries of the world become modernized, their birth rates fall. And despite population increases in recent decades, agricultural output has gone up even faster—on less farmland!

Now, global warming is the *cause du jour*. Environmentalists, politicians, clergy, doctors, actors, musicians, and representatives of probably every other profession have all spoken out about the danger that global warming poses to both humanity and the Earth.

That mankind inadvertently influences the weather is true, at least to some extent. It would be surprising indeed if the climate system did not notice that six billion people live on the Earth. *Everything* influences the weather. Why should it be any different for humans? A forest changes the weather from what it would otherwise be if the forest did not exist. The same goes for lakes and oceans, rivers, plains, and mountains. We might have a fond attachment to deserts, but think objectively about what they really are: vast stretches of nearly dead land.

The Romantic notion that nature untouched by man is "pristine" is a philosophic, even religious, point of view. Why do we give nature a pass, but not ourselves? I find such attitudes fundamentally anti-human, and certainly not scientific. As long as we keep being told, explicitly in news stories, or implicitly through movie themes, that we are the enemies of the environment, then we will be too meek to stand up for ourselves and our right to use nature for our own purposes. I believe that the only rights that the natural world has are those conferred upon it by humans.

Once we elevate the concerns of nature above those of people, we abdicate our authority to do the things that are necessary to improve the human condition. Yet you seldom hear this point of view being advanced. It is considered politically incorrect, anthropocentric, arrogant, or even worse—capitalistic.

I am part of the relatively small, infamous minority of climate researchers known as global warming "skeptics." Despite the oft-repeated claims of our detractors, it is not true that we do not believe in global warming. Al Gore has grown fond of calling us "global warming deniers," apparently hoping to confuse the public through propaganda, knowing full well that none of us deny that global warming has taken place. What we are skeptical of is the theory that all (or even most) of global warming is caused by mankind, or that we understand the climate system and our future technological state well enough to make predictions of global warming in the next fifty to one hundred years, or that we need to reduce fossil fuel use now.

There are two themes in environmentalist rhetoric that seek to discredit us so-called skeptics on global warming issues. The first is that corporations with lots of wealth buy influence from skeptics, and therefore we can't be trusted. The second is that skeptics use scientific disinformation in their attempts to undermine the scientific consensus that global warming is real.

On the first point, contrary to what most would expect, the financial incentive for individual scientists to speak out on global warming is on the side of the global warming alarmists. While private industry would seem to have the most money available to "buy" opinions, big corporations tend to shy away from that kind of influence. For instance, in my case I have never been approached by any energy company seeking to pay me for any service. I wrote "skeptical" articles and book chapters, for no pay, for thirteen years before a science and technology website, TechCentralStation.com, offered to pay me to write articles about the latest newsworthy events that involved global warming.

While I have given talks to organizations which are partly funded by "Big Oil," I have also given similar talks to state environmental organizations. Left-leaning websites like ExxonSecrets.org mention only the former in their attempts to make it look like we global warming optimists are simply shills for big business. This guilt-by-association tactic helps them avoid having to address our arguments based on science.

Corporations recognize the need for government-sponsored research to help answer scientific questions since that research is presumably unbiased. But as we shall see, the governmental funding of researchers is definitely biased toward work that demonstrates that global warming is a threat, since this helps to maintain research programs at NASA, the National Oceanic and Atmospheric Administration, the EPA, and the Department of Energy.

In contrast, philanthropic foundations with leftist boards of directors routinely give money to alarmist causes. For instance, $500,000 no-strings-attached grants have been awarded by the MacArthur Foundation to climate researchers who speak out publicly for the global warming threat. James Hansen, the director of NASA's Goddard Institute for Space Studies, received a $250,000 grant from a foundation headed by John Kerry's wife, Teresa Heinz Kerry. That Hansen publicly endorsed John Kerry for president in 2004 is claimed to be an unrelated coincidence.

There are no such conservatively funded monetary awards that I am aware of. And based upon its historical record, you can bet that a Nobel Prize will never be awarded to the scientist who ever demonstrates that global warming is not the huge threat to mankind that it is advertised to be.

While there are a number of pro-free market organizations that receive funds from big corporations, the dollar amounts pale in comparison to the budgets of environmental organizations. By far the largest supporter of environmental groups and climate researchers is the federal government, with your tax dollars. And the dirty little secret is that many environmental organizations are also funded by Big Oil.

For many years now, well over $100 million a year has been flowing from the federal government to environmental lobby groups. The federal government routinely funds so-called non-governmental organizations (NGOs) that turn right around and lobby the government to support environmental causes that the NGOs depend upon for their survival. Yes, I know this seemingly incestuous relationship would be inconsistent with the high regard you have for politicians, but trust me, it is true.

The environmental movement is indeed a huge financial machine with all the power and influence that comes with money. What happens to this machine if interest in environmentalism wanes? At least for-profit corporations offer goods and services that people will continue to need. In contrast, without a constant supply of environmental scares, environmental organizations will simply die.

I am *not* claiming that environmental organizations shouldn't be funded. I *am* saying that they should not be throwing stones while operating out of glass buildings.

The second accusation about global warming skeptics is that we sow scientific disinformation to undermine the scientific consensus that "global warming is real." I would call *that* disinformation. Every scientist-skeptic I know believes that global warming is real. Instead, the central questions being debated are: How much of the Earth's current warmth is the result of natural processes versus the activities of mankind? How bad will global warming be in the future? And maybe most importantly, what can and should be done about it?

While science can give us some useful information on the threat of global warming, it has nothing to say about our response to it. Science is values-neutral and policy-neutral. Instead, what should be done about global warming comes from people's belief systems: their opinions of the proper role of government, understanding of economics, and even their religious faith and worldview.

Like previous authors, I could have written a book on the dry, scientific evidence for and against global warming theory, and what scientists currently believe about the threat that global warming poses to mankind. And this book does include explanations of how hurricanes, tornadoes, and less newsworthy weather events relate to global warming. But as scientific understanding changes, such books can quickly become outdated.

While I will refer to a few important works that support my views, I will avoid detailed listings of scientific findings, pro or con. These give the impression that stacks of evidence in the pro-warming or anti-warming pile determine who wins the scientific debate. And while it is true that more scientific findings are supportive of global warming theory than those that aren't, we will see that this is largely the result of the research funding deck being stacked against us skeptics.

Rather than discussing the latest global warming research and what it means, I will instead address the overriding issues and concepts that will not soon change in the scientific debate. I will describe why I believe that the Earth's climate system is not nearly as fragile as most computerized climate models tell us it is, and what amounts to the climate system's thermostatic control mechanism.

An informed public is vital during this age of political pandering to constituent's views. The mainstream news media not only decides what you should know, but tells you what you should think about it. They uncritically accept every environmental scare. In their imaginary world, environmental regulations have no downside, and we can have all benefits with no risks.

This book is one small effort to help balance those influences in the global warming arena. I am now convinced that currently proposed global warming policies will actually do more harm than good—to both humanity and the environment. I will explain, in simple terms, why so many scientists believe that manmade global warming is a dangerous threat, and why I believe that they are wrong.

I will explain why the theory of manmade global warming will always remain just a theory, despite increasing numbers of people who are trying very hard to convince you it is fact. The emotional attachment that these people have to catastrophic global warming can be traced to a variety of self-interests—careers, political and social policies, philosophies and religious beliefs—all masquerading as science.

And since policy decisions are usually economic decisions, unless we understand basic economic principles, it is impossible for us to have any meaningful opinions on what should be done about global warming. Even though environmentalists are insisting that we do something now about global warming, I will demonstrate why the unintended negative consequences of such a view might well do more harm than good. If you read only one chapter in this book, I suggest Chapter 6 (It's Economics, Stupid) —it really is that important.

So, while we are waiting for the predicted meltdown of planet Earth, I would like to guide you through not only the science issues, but also the philosophical, economic, political, and even religious elements that cannot be separated from how we view the global warming problem.

Critics of this book will say that my treatment of global warming is obviously biased. And they are right. I have studied the issues enough to have developed some very strong biases on the subject. But it is not a question of whether bias exists—for we are all biased. It is a question of which bias is the best bias to be biased with.

Run for your life! Run for your life!
The ice age is coming!!

Chapter 1: Global Warming Hysteria

It was a dark and stormy night; the rain fell in torrents—except at occasional intervals, when it was checked by a violent gust of wind which swept up the streets (for it is in London that our scene lies), rattling along the house-tops, and fiercely agitating the scanty flame of the lamps that struggled against the darkness. Had Jack known that such meteorological chaos, long predicted by climate experts and his favorite movie stars, would have ensued after his purchase of that petrol-guzzling behemoth, he would not so quickly have given in to the siren calls of the TV commercials that had relentlessly nagged him into buying his Hummer—an acquisition that would trouble his heavy heart, a good heart, right up to this very moment.

IN CASE YOU have not noticed, all natural disasters are now caused by global warming. Tsunamis, hurricanes, tornadoes, heat waves, and snowstorms are all being blamed on mankind's use of fossil fuels. The latest flood and drought were both caused by global warming. Al Gore's movie and book *An Inconvenient Truth* make it sound like chunks of glaciers breaking off and falling into the ocean, arctic sea ice melting, and major hurricanes striking the United States never happened before global warming ... and if we would just stop the warming, these things wouldn't happen ever again. Even the Alaskan Inuit's "traditional" way of life is now threatened by warmer temperatures in the Arctic. It seems their snowmobiles are falling through the ice more frequently these days.

Apparently, global warming theory is so powerful and flexible that it can explain everything, from failed crops, to flooded homes, to shrinking polar bear populations, and, as recently reported, even shrinking polar bear testicles. Warmer winters? Evidence of global warming. Colder winters? Also evidence of global warming. The theory of manmade global warming has been elevated to a physical law, proven beyond any doubt, and it supposedly now gives us a unified way to explain any change we see in nature.

But no matter whether manmade global warming is a serious problem, an overblown fear, or even nonexistent, it does provide a source of some excitement and entertainment in our lives. Let's look at some of the ways in which we have been entertaining ourselves with global warming.

RECORD WEATHER EVENTS

It appears to be part of human nature to blame severe weather events on something done by mankind, whether it is global warming, or the Russians' weather control machine that a South Dakota TV weatherman blamed the 2005 hurricanes on. But one of the features of extreme weather events that most people don't understand is that abnormal weather is ... well ... normal.

Let's take record high or low temperatures as an example. For a weather observation site that has existed for one hundred years (and there aren't many of those), daily record high temperatures should, on average, occur three or four times a year. The first year the station existed, every day experienced a record. This is because the high temperature was higher (and lower, too) than it was ever measured before on that date ... which was never. Then, in the second year, one-half of the days would have record high temperatures, while all the others would be record low temperatures. Only in the third year would it even be possible for a temperature not to set a new record every day.

You get the idea. It is normal, even expected, to have record weather occurrences on occasion. Yet we seem surprised when they happen. They also tend to occur in clusters. Some regions will have record temperatures for days in a row. One hundred or more cities and towns might set records on a given day during a heat wave. And while you might have been led to believe that the all-time record high temperatures in the United States were set in the last ten years or so, the truth is that the decade with the largest number of all-time state record high temperatures was the 1930s.

And what happened in previous centuries? Was the weather back then really that different from today? No one really knows for sure. If it was different, that wasn't because of humans. While there is some anecdotal evidence about grapes growing in England, or the Thames River freezing over completely in winter in centuries past, these events might well have had no connection to global warmth or cold. The Earth is a pretty big place, and since three-quarters of it is ocean, it still remains a little difficult to measure what is happening everywhere.

One thing we do know is that, historically, warm weather has been better for humans than cold weather. Around 1000 A.D., warm climate conditions called the Medieval Warm Period, or Medieval Optimum, existed. Humanity prospered during this time, presumably because they weren't buried under a mile-thick layer of ice like they were during the "Less-Than-Optimum Ice Age." Note that climatologists call that warm period the "Medieval Optimum," not the "Medieval Global Warming Disaster."

Then there was the widespread fear back in the 1970s that the slight cooling trend we had been experiencing since the warm 1940s was the start of the new Ice Age. People instinctively knew that cold was bad (unless you operate a ski resort). But now warmer is also bad. Apparently, the exact temperature we were at in 1980 was the temperature we are supposed to be at. It is perfect, undefiled, and natural. Never mind that "perfect" for many of us ranges from below 0° Fahrenheit in the winter to 90° Fahrenheit in the summer. Those, presumably, are the temperatures ordained by Mother Nature, and we shouldn't even think about touching the thermostat.

As the following chart of global temperatures between 1850 and 2005 shows, however, globally averaged temperatures can change substantially for entirely natural reasons. Most of the warming up until 1940 could not have been caused by mankind simply because we had not emitted very much in the way of greenhouse gases before that time.

And when it comes to actually knowing what global temperatures were before about 1850, we simply do not have the meas-

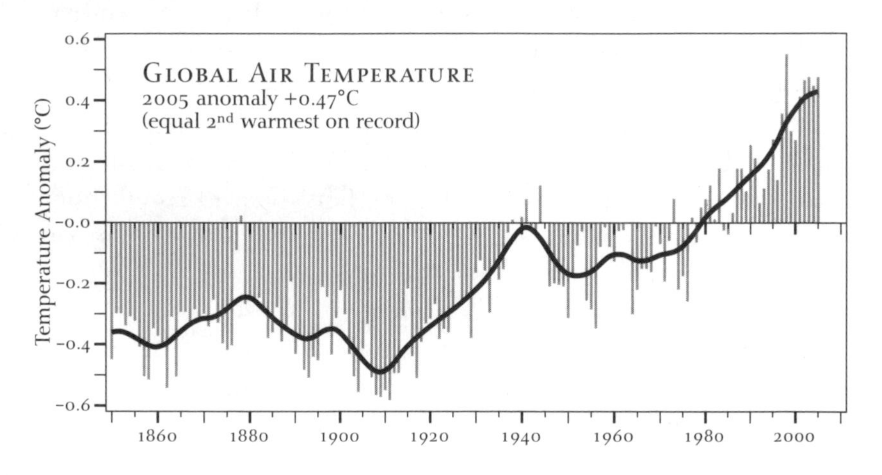

urements available to say much of anything of scientific value.

Nevertheless, some scientists won't let a little problem like a lack of measurements stop them. A number of scientists, apparently frustrated historians, have created a discipline called "paleoclimatology." This is where scientists look at tree rings or ice core layers and magically divine the historical temperature record.

While normal people would call such an interpretation questionable, the researchers prefer to call it science. With just a few key assumptions (which are then immediately forgotten), these paleoclimatologists can tell us what the weather was like during past centuries or millennia. Since they do not have to deal with the inconvenience of actual temperature measurements to verify their methods, their results are often treated as gospel. There's an old joke in science that if you want perfect measurements, take them only once. Then there's no disagreement.

I personally do not put much faith in paleoclimate studies. Since scientists can't even agree on the accuracy of actual thermometer-measured temperatures over the last hundred years, I find claims that we can discern ancient temperatures based upon the tree-ring spacing of a Bristlecone Pine growing at 9,000 feet elevation in a remote corner of Colorado to be a little dubious.

A National Academy of Sciences review panel in 2006 addressed an ongoing flap about whether the Earth is warmer now than anytime in the last 1,000 years. The infamous "hockey stick" curve of global temperatures published by paleoclimatologists in 1998 made a huge splash because it downplayed the warmth of the Medieval Warm Period, thus elevating the late twentieth-century warmth to record status. Without explicitly scolding the hockey stick inventors for using questionable statistical techniques to make their warmest-in-a-thousand-years claim, the NAS panel finally agreed that about all one could say with considerable confidence is that the Earth is warmer now than anytime in the last 400 years.

The media exclaimed, "Oh, no! We are warmer now than anytime in the last 400 years!," apparently not realizing that this was a pretty big downgrade from their previously reported exclamation, "Oh, no! We are warmer now than anytime in the last 1,000 years!" Since 350 of those 400 years were during the "Little Ice Age," I would call our current warmth pretty good news.

Yet, even though our present period of warmth might be more beneficial than harmful, most journalists can't ever bring themselves to report anything in a positive light. What a depressing job.

Another human tendency that influences our perception of global warming is the belief that, if we happen to be experiencing a drought, the whole world must be drought-stricken. If this is the warmest year our town or city has ever recorded, then the whole world must be sweltering. Any unusual local weather event is given global significance. Maybe you have heard of "weather patterns"? Well, they really do exist, and they influence relatively small portions of the Earth. Of course, we scientists don't call them "weather patterns," because that sounds too unprofessional. We call them "regional climate anomalies," a more technical term which shows how intelligent we are.

Even the United States, as big as it is, represents only a little over 2 percent of the surface area of the Earth. I once computed the correlation between United States temperatures and globally averaged temperatures from the satellite data analyses that John Christy and I originally developed and continue to maintain. (Satellites, by the way, provide our only way of measuring the whole globe.) The resulting correlation was just about zero. No relationship. Even averaged over the entire United States, heat waves or cold snaps were unrelated to globally averaged conditions.

And that was for the whole United States, not just a little burg like Podunk, Michigan (which, by the way, is a real town). When one region is experiencing unusually hot weather, it is almost always accompanied by another area, typically a thousand miles or so away, that is having unusually cold weather. For instance, in the United States, if the east coast is having a heat wave, you can usually count on California being unseasonably cool, and vice versa. It doesn't have any global significance. Well, okay, it does have a little over 2 percent global significance.

So when Al Gore gave a global warming speech in January 2004 on one of the coldest days ever in New York City, one could have been tempted to attribute global significance to the event, rather than just irony. But that cold day really didn't mean that the whole Earth is getting colder. Nevertheless, you can bet that when Earth Day occurs during a heat wave in New York City, the event speakers will tie that to global warming.

Another issue that colors how we view extreme weather events is that we tend to put current weather into the context of our own, rather short, lifetimes. If we never experienced something before, we are tempted to conclude that it never happened before at all. And even within our lifetimes, our memory of past events that really did occur is often not very good. The best example of this is the Great Alien Invasion of 1984. Absolutely no one that I ask about this historic event can even recall it happening. A brilliant illustration of my point: the public has a short memory.

And how many "storms of the century" did we have in the 1980s and 1990s? As I recall, by 2001 or so, we had already experienced what was called the "Storm of the 21st Century." Apparently, someone entered NASA's secret weather time machine, and checked out what will happen over the next ninety-nine years. Um ... forget I told you about the time machine, you aren't supposed to know about it. Oh, never mind, you'll forget about it anyway.

For something blatantly ridiculous, let's look at a non-weather event that some people nevertheless connected to the weather. The mega-tsunami of December 26, 2004 in Indonesia certainly was unprecedented in recent human memory. Unbelievably (or maybe predictably), a few experts and politicians actually blamed the tsunami on global warming.

Well, tsunami waves are triggered by earthquakes under the bottom of the sea. The earthquakes, in turn, are caused by the tectonic plates that make up the Earth's crust slowly grinding against each other. Given the enormous forces deep in the Earth involved in earthquake formation, any change in the concentration of a minor atmospheric constituent like carbon dioxide would go totally unnoticed by the Earth's crust.

The Earth's crust and mantle simply don't care whether the atmospheric carbon dioxide concentration is 250 parts per million (sometime in the distant past) or 500 parts per million (predicted for late in this century). It probably doesn't even care whether there is an atmosphere or not. I tried to invent a more absurd reason than global warming for the cause of the Indonesian tsunami, but I couldn't think of one. Yet since the supposed connection to global warming was reported in the news, many people will believe it.

While most experts didn't blame the tsunami on global

warming, it was at least suggested that the tsunami showed how vulnerable we are to global warming. But this historic tsunami event should make us fear global warming less, not more. The tsunami puts global warming into proper perspective for us. Coastal residents have been told that global warming will cause sea levels to rise by several inches in the coming decades. Then along came a tsunami, growing to as much as twenty feet tall in a matter of seconds, making a possible gradual sea level rise of a few inches all but superfluous, lost in the noise, irrelevant.

The difference in magnitudes between these two extremes is astounding. A gradual sea level rise can be adapted to. Seawalls and levees can be built up; the next crop of new buildings can be on slightly higher foundations. But an earthquake under the sea floor is totally unpredictable. The only way to avoid tsunamis is to not live so close to sea level. Like that is going to happen.

The inescapable reality is that people who live in coastal regions are at risk of natural disasters visiting them from the ocean, just as those who live in earthquake zones are at risk, and those who live where tornadoes occur. There are very few places on Earth that are not visited by some sort of natural calamity—and no one would want to live in those places anyway for fear of dying from boredom.

HURRICANES

Louisiana and Mississippi are still reeling from the multiple major hurricanes that hit the Gulf of Mexico coast in 2005. After these weather disasters, Barbra Streisand warned us that "We are in a global warming emergency state, and these storms are going to become more frequent, more intense." Maybe I would take her more seriously if I liked her songs more, since that is apparently why she thinks anyone would care about her opinions on global warming. It would also help her credibility if she, say, happened to be a hurricane or climate expert.

Another esteemed hurricane expert, Robert F. Kennedy, Jr.,

wrote after Hurricane Katrina devastated southern Mississippi and southeastern Louisiana that this was the Earth's punishment on Mississippi's Governor, Haley Barbour, for supporting the use of fossil fuels. Until Kennedy had informed us of this, I was completely unaware that the Earth was so vindictive. I'm sure Mr. Kennedy makes certain that he doesn't use fossil fuels.

It might be hard for reporters to believe, but the fact is that major hurricanes have always been a threat to the United States. Multiple major hurricanes have hit the United States during its history, even before global warming was at fault for everything. Hurricane experts at the National Hurricane Center have been warning for decades that we were in a lull of hurricane activity, and that it was only a matter of time before the natural thirty- to forty-year cycle in hurricane activity went into its more active phase once again.

The dominant reason for the increased hurricane threat is well known to be a result of so many people flocking to the coasts in recent years. Coastal construction has skyrocketed, and the more buildings there are, the more targets there are for hurricanes. This is analogous to trailer parks in tornado-prone areas, which some meteorologists call "tornado bait."

Despite the very real and large natural variability in the number of hurricanes, there have been a couple of studies that have suggested that global warming could at least cause some average intensification of these storms, by several percent. Two more studies have made a connection between recent warmer sea surface temperatures and tropical cyclone strength.

But an undue emphasis on this possibility distracts us from the bigger issue: major hurricanes always have, and always will, hit the United States. The public needs to be prepared for them, with or without a small average increase in intensity. If you live in a vulnerable coastal area, you cannot expect the Mayor, the Governor, Congress, FEMA, or the President to save you.

Even the experts say that you probably can't blame the most recent upswing in hurricane activity (2004 and 2005 at this writing) on manmade global warming. Max Mayfield, the current director of the National Hurricane Center; Neil Frank, the previous director of the Hurricane Center; Chris Landsea, the leading hurricane researcher; Bill Gray, the famous seasonal hurricane forecaster: These experts doubt that manmade global warming is to blame. But the former next President of the United States, Al Gore, disagrees, and has written a book and made a movie to help convince you that he is right and they are wrong.

After a busy 2004 and record 2005 hurricane season, 2006 was expected to be very busy as well. To everyone's surprise, though, Mother Nature provided a below-average hurricane season. In fact, right at the peak of the season, there were no tropical cyclones at all, either in the Atlantic or the Pacific. Obviously, this unusual event must have been caused by global warming, too.

Like any other kind of climate forecasting, forecasting the hurricane season months in advance is a very risky business. What saves these long-range prognosticators from the embarrassment regular weather forecasters have to endure on a daily basis is that, by the time the climate forecaster is proved wrong, everyone has forgotten about his forecast anyway. Instead, we are busy worrying about the new long-range forecast.

The forecasters at the National Hurricane Center (NHC) in Miami readily admit that there is little skill in forecasting tropical cyclone intensity changes, even hours in advance. But the public still expects them to forecast something about what a given hurricane or tropical storm will do in the coming hours and days. The necessity to produce six-hourly forecasts for tropical cyclones, combined with the stubborn habit of those cyclones to ignore the forecasts, can lead to some frustrating times for hurricane forecasters. The following rather humorous series of NHC headlines appeared on advisories for what eventually became Hurricane Florence that hit Bermuda in September of 2006:

Advisory #8, 11 am Tue Sep 05 2006:
...The sixth named storm of the season develops over the central tropical atlantic ocean ...

Advisory #9, 5 pm Tue Sep 05 2006:
...Florence getting a little stronger over the open atlantic ...

Advisory #10, 11 pm Tue Sep 05 2006:
... Florence expected to strengthen ...

Advisory #11, 5 am Wed Sep 06 2006:
...Florence remains over the open waters of the central Atlantic ...

Advisory #12, 11 am Wed Sep 06 2006:
...Florence a little stronger ...

Advisory #13, 5 pm Wed Sep 06 2006:
...Florence getting better organized ... should strengthen soon ...

Advisory #14, 11 am Wed Sep 06 2006:
...Florence has not strengthened yet ...

Advisory #15, 5 am Thu Sep 07 2006:
...Florence holding steady and moving toward the west-northwest ...

Advisory #16, 11 am Thu Sep 07 2006:
... Florence moving west-northwestward over the open Atlantic ...

Advisory #17, 5 pm Thu Sep 07 2006:
... Florence geting a little better organized over the open Atlantic ocean ...

Advisory #18, 11 pm Thu Sep 07 2006:
... Florence remains large but refuses to strengthen ...

Advisory #19, 5 am Fri Sep 08 2006:
...Florence shows little change...

Advisory #20, 11 am Fri Sep 08 2006:
...Florence appears to be ready to strengthen...

Advisory #21, 5 pm Fri Sep 08 2006:
...Florence has not strengthened yet ... but it could tonight...

Advisory #22, 11 pm Fri Sep 08 2006:
...Florence intensifying...

Advisory #23, 5 am Sat Sep 09 2006:
...Florence moving west-northwestward and gradually becoming better organized...

Advisory #24, 11 am Sat Sep 09 2006:
...Florence continuing to show signs of getting better organized...

...Tropical storm warning issued for Bermuda...

Advisory #25, 5 pm Sat Sep 09 2006:
...Florence weakens slightly but is expected to strengthen again on Sunday as it approaches Bermuda...

(By this time, half of the people in Bermuda had bitten all their nails off, and the other half had gone back to the beach.)

Advisory #26, 11 pm Sat Sep 09 2006:
...Florence turns toward the north-northwest with no change in intensity...

And finally (drum roll please)...

Advisory #26a, 2 am Sun Sep 10 2006:
...Florence reaches hurricane strength...

Please understand that I hold the forecasters at NHC in very high regard. They take their jobs seriously, and they do the best they can. But unfortunately, our understanding of what causes a specific hurricane to form, strengthen, or weaken is just not very good yet. Warmer sea surface temperatures can, on average, lead to stronger hurricanes. But if warming water temperatures are also accompanied by an increase in wind shear (a change in wind direction or speed with height), we could actually see fewer and weaker hurricanes accompanying warming. Sea surface temperatures are only part of the story.

Another influence on the number of hurricanes is the presence or absence of "seedlings." Most of the tropical cyclones that form in the Atlantic basin can be traced back to easterly waves which travel westward off of sub-Saharan Africa. These provide a favorable environment for a tropical depression to form. Any change in drought or rainfall conditions over Africa can alter the strength and number of these disturbances that Africa sends our way, and so influence Atlantic hurricane activity. Water temperatures, wind shear, the presence of African easterly waves, and the weather patterns over Africa—these are some of the variables that cause so much year-to-year variability in hurricane activity.

Finally, believe it or not, sea surface temperatures themselves are not subject to just the activities of mankind. Research published in late 2006 showed that the globally averaged temperature of the upper ocean cooled so fast between 2003 and 2005 that over 20 percent of the warming that had occurred over the previous forty-eight years was cancelled out in just those two years. And the reason why, you ask? No one knows.

The bottom line is that focusing too much on whether or not global warming will make hurricanes a little stronger diverts our attention from a bigger issue. Category 4 and 5 hurricanes have always existed, always will exist, and if you build too close to sea level in a hurricane-prone region, it is only a matter of time before one comes to your town, too.

———

TORNADOES

Every time there is a tornado outbreak, you can count on someone bringing up global warming as a possible cause. For instance, the tornado season of 2004 saw record tornado activity. But unlike the case for hurricanes, there is no research showing changes in tornado events with global warming. Tornadoes require a severe thunderstorm embedded in the proper wind shear conditions, usually near the boundary (front) between two air masses of different temperature. It is simply unknown how these conditions might change with global warming. As with hurricanes, the year-to-year variability can be astonishing—but it is entirely natural.

Has there been a recent upswing in tornado activity that could possibly be blamed on global warming? Well, there has been an increase in the number of tornadoes *reported* over the last fifty years. Unfortunately, it is a little difficult to construct a believable long-term record of tornado occurrences. In the early years there were fewer people, spread over a smaller geographic area, with even fewer cameras than we have today.

Now there are people living and traveling more widely, there are more video cameras, and most of the country is covered with Doppler radars capable of revealing tornadic thunderstorms. A tornado can't have any privacy anymore, as it is sure to be observed by someone. Thus, it is well known by experts that the reported increase in total tornado occurrences in the United States over the last fifty years has been heavily contaminated by this effect.

A better idea of whether there has been a real upward trend in tornadoes in recent decades comes from the reported number of strong to violent tornadoes. This trend has remained flat. Because violent tornadoes are longer lived, cover a larger area, and leave a bigger mess behind, they are much less likely to go unnoticed, even in remote areas.

Mainstream Media Madness

No one does a better job at keeping you misinformed on environmental issues than the mainstream news media, which increasingly tries to entertain you, and the entertainment industry, which increasingly tries to tell you what to believe about the newsworthy events. A large part of the public's concern about the environment can be traced to editorial bias that exists in the major media sources.

Journalists are no longer interested in keeping you informed. Instead, they are out to change the world. Ever since the Watergate scandal propelled the *Washington Post* reporters Bob Woodward and Carl Bernstein to fame during Richard Nixon's presidency, reporters have lusted after the big scoop that will get them a Pulitzer Prize. Journalists today are falling all over themselves to convince you of how serious global warming will be. If the prize is ever given for climate change reporting, it will have to be shared by 1,735 journalists, all of whom have broken the story that a global warming Armageddon is coming.

The very fact that news is, almost by definition, something startlingly different from normal means that there is plenty of room for both journalistic and scientific bias to creep in. After all, just like the scientist who wants to be the one to make a new and startling discovery and be awarded a Nobel Prize, the journalist wants to break the Big Story and receive a Pulitzer Prize.

The media can always find an expert who is willing to provide some juicy quotes regarding our imminent environmental doom. Usually, there is a grain of truth to the story, which helps sell the idea. Like a science fiction novel, a somewhat plausible weather disaster tale captures our imagination, and we consider the possibility of global catastrophe. And some of the catastrophic events that are predicted are indeed possible, or at least not impossible. Catastrophic global warming—say, by 10° Fahrenheit or more over the next century—cannot be ruled out with 100 percent certainty. Of course, neither can the next extraterrestrial invasion of Earth.

But theoretical possibilities reported by the media are far from competent scientific predictions of the future. The bias contained in all of these gloom-and-doom news stories has a huge influence on how we perceive the health of the Earth and our effect on it. We scientists routinely encounter reporters who ignore the uncertainties we voice about global warming when they write their articles and news reports. Sometimes an article will be fairly balanced, but that is the exception.

Few reporters are willing to push a story on their editor that says that future global warming could be fairly benign. They are much more interested in gloom and doom. A scientist can spend twenty minutes describing new and important research, but if it can't be expressed in simple, alarmist language, you can usually forget about a reporter using it. It has reached the point where the minimum amount of necessary alarm amounts to something like, "we have only ten years left to avert catastrophic global warming." A reporter will probably run with that.

After all, which story will most likely find its way into a newspaper: "Warming to Wipe out Half of Humanity," or "Scientists Predict Little Warming"? It goes without saying that, in science, if you want to keep getting funded, you should find something Earth-shaking. And if you want to get your name in the newspaper, give a reporter some material that gives him hope of breaking the big story.

The media alarmism, even hysteria, over the supposed threats posed by global warming might have reached a new high with the April 3, 2006 issue of *Time* magazine. The cover story was entitled "Be Afraid, Be Very Afraid." And if you didn't know any better, after reading the articles in that issue you definitely would be afraid.

Now, I'm sure that the journalists who wrote these *Time* articles were simply trying to provide balanced information on an important issue so that the public can make informed decisions.

Ha-ha! That was a good one. The *Time* articles actually scored a big fat zero in the objectivity department. Phrases such as "the climate is crashing," "the crisis is upon us," and "nature has finally got a bellyful of us" could be found without even turning to the second page of *Time*'s twenty-six pages of predicted disaster.

Apparently *Time* did not see the irony of their articles being interspersed with ads for new SUVs.

Every possible storm was pointed to as another piece of evidence of global warming. The "atmospheric bomb that was Cyclone Larry" had just struck Australia, supposedly driving one more symbolic nail in the coffin of global warming skepticism. But then, within a few days it was realized that Larry was probably only a weak category 4, not a category 5. Much stronger cyclones have occurred in the region in the past. Since Australia doesn't fly research aircraft into these storms to measure their strength like the United States does, Australian meteorologists usually don't have a very good estimate of how strong they are. As a result, forecasters err on the side of safety, and over-warn the public.

And never mind that many of the weather events that *Time* pointed to as evidence for global warming (floods, droughts, storms, crop failures) are the same as those that were blamed on the approaching Ice Age in a *Time* issue back in June of 1974. To be fair, *Time*'s main competitor *Newsweek* also performed the same about-face, switching from a predicted Ice Age disaster to a global warming disaster.

It is obvious that the media is already biased towards pro-warming stories. I hear it in the questions that reporters ask me and in the tone of their voices. They deceive you about what direction they claim to be taking a story, sounding sympathetic to your views. Then, they end up putting a dizzying spin on the final product. If journalists and their editors really like doing these kinds of stories, why don't they just work for *National Enquirer*?

I suppose I should be sympathetic to the plight of journalists. They have to cover a wide range of issues on a daily basis, reporting on concepts that they do not have time to understand fully. So some level of inaccuracy in their reporting is to be expected. But the public definitely gets a biased view if they rely on the mainstream news media to keep informed about global warming. Nuances and uncertainties get lost, and the scientific claims that appear in print, after being rephrased by the reporter for clarity, are often misleading or outright wrong.

There is a perverse incongruity that attends mainstream media stories of predicted environmental disaster, a warp in the journalistic space-time continuum that they seem entirely comfortable with. It feels like a TV news flash that interrupts current programming—"World Ends! Details at eleven."

It's like the Jim Carrey movie *The Truman Show*, in which Carrey plays a man whose entire life has been part of an orchestrated play acted out in a small town isolated from the rest of the world. Secretly broadcast around the world on TV, humanity has been mesmerized by his life story since he was born.

When Truman finally discovers the truth and the last show ends with Truman escaping his artificial bubble of existence, the whole world watches in stunned silence—for about ten seconds. Then they begin flipping through the channels to see what else is on TV. In the same way, we have become accustomed to breathless reports of weather and climate disasters. On the positive side, I have found that they at least provide a well-deserved ten-second break for the thumb while channel surfing.

While the reporting of environmental cataclysm is convincing you that humanity is already past the point of no return, you are also treated to the latest gossip about Britney Spears. We are told that we might have only a few years left to repent and get rid of all of our gas-guzzling cars and trucks, and then we see a commercial for the latest powerful SUV that has hit the market. Call me crazy, but there is something more than a little schizophrenic about environmental reporting.

No wonder our children are having nightmares. Global warming and Britney Spears mixed together in a dream could cause anyone to wake up in a sweat.

You would think that the scientific journals would be better than newspapers, since that is where the full peer-reviewed research report gets published, in the researcher's own words. And for the most part, the journal system of scientific reporting works pretty well. But there are a few "gray literature" science publications which are little more than a science tabloids, and yet they command the greatest amount of attention from reporters.

The two most famous of these publications are *Nature* and *Science*. I have experienced newspaper-reporter-style pro-global warming bias especially on the editorial board of *Science*, as have other scientists I have talked to. In all fairness, *Science* policy does admit that they are only interested in reports of broad interest to the science-savvy public. Unfortunately, this policy results in an automatic bias in what papers get published. Any research report on the latest threat to the environment automatically has a foot in the door. Any research that says the threat is not serious might as well look elsewhere for publication.

Since these pop-science journals are considered to contain the most newsworthy results rather than the most carefully performed science, they will typically publish only those articles that can be expected to make a big splash. If the editors want to publish a paper that has a conclusion they like, they seem to know whom to send the manuscript to for a good review.

Most of the papers published in this scientific gray literature are required to be very brief, and so do not have sufficient detail in them to allow the reviewers of those papers to be convinced that the research was carried out thoroughly and carefully. The reviewers simply have to take the author's word for it.

Further damaging the scientific value of these publications is their reliance on relatively few reviewers of the manuscripts. At times, these reviewers are not even that familiar with the line of research. I have noticed *Science*'s reliance on one particular scientist for reviews who seemingly never saw a pro-global warming paper he didn't like, or an anti-global warming paper that he did like.

Finally, if a newsworthy article that is published is found to by other researchers to be flawed, it is particularly difficult to get those contrary results published in the same journal. The challenge is similar to that of getting a newspaper to print a retraction. I suppose that if it happened too often the editorial staff of the magazine might look incompetent, so there is a disincentive to publish contrary results.

And the problems don't even end there. The editorial bias at pop-science magazines then tends to have a trickle-down effect on new research funding. Peer-reviewed scientific publications, being the ultimate goal of scientific research, help a researcher to get more funding from government agencies. Government managers seem to think that *Science* and *Nature* are the premier scientific journals, and so publications in them carry greater weight than papers published in the mainline scientific journals. Thus, it is possible for a biased line of research to get perpetuated once it has been kick-started with the publication of the initial research effort.

A famous example of bad science that found its way into one of these gray literature magazines involved research concerning the behavior of the climate system. The researcher who submitted the paper for publication was trying to show evidence for a strong relationship between two variables, let's call them X and Y, on a graph. Since there wasn't much of a relationship (which apparently didn't fit his preconceived notions), he discovered that a much stronger relationship could be obtained if he plotted X versus X-minus-Y. What he, and the incompetent reviewer(s), failed to realize is that this has no physical significance. Even if two variables are totally unrelated to each other, X still will always show a good relationship with (X-Y), for the simple reason that the variable X is contained in both!

The media hype and hysteria over the dangers of global warming seem to only grow with time. Each subsequent news story has to be more alarming than those that have preceded it. At first, the news was that global warming would be gradual, stretched out over many decades.

Then the stories of "tipping points" started, wherein our tinkering with the climate system would cause sudden climate shifts. For instance, the upper-ocean Gulf Stream or the deep-ocean thermohaline circulation would suddenly shut down, causing a mini-Ice Age to occur in Europe. Then we were told that we have only ten years left to avert a climate catastrophe, and that we might be warmer now than anytime in the last million years.

Finally, we are now being told that the catastrophic effects of global warming are here and now. ABC News online even started asking for the public's stories about how global warming has changed their lives. The facts don't matter anymore—what is important are people's perceptions. Global warming is serious simply because we think it is serious. Facts are facts, but perception is reality. I'd like to get a bumper sticker that says "Imagine Global Cooling," but I'm afraid that the point would be lost on most people.

One of the casualties of this incessant onslaught of media-enhanced doomsday predictions is that about one-half of the public has started to believe that scientists really are worried about all of these things. Our children are literally frightened. Many folks are becoming less discerning about what they read. It has reached the point where the outlandish environmental predictions that appear in the news are stranger than fiction. They have become more fantastic than what someone with a perversely fertile imagination could dream up.

And being one of those perversely imaginative people, this realization gave me an idea.

A Web-based Trip to the Twilight Zone

I became intrigued not only by some of the wild claims that were appearing in the news about various environmental ills, but also by the willingness of so many people to believe those claims. My curiosity about this phenomenon led me to start a website called EcoEnquirer.com. I made up and posted fake, satirical environmental news stories. In each story I included enough hints for the discerning reader to realize they were reading fiction.

One of the cool things about running a website is that you get to see what people are saying about your site on internet discussion forums. To my surprise, at least one half of the people who commented about my articles actually believed them to be true. Some felt a little foolish after other people pointed out that what they were reading was satire. The most common comment I read was to the effect that environmental news stories had become so crazy lately that it was impossible to tell satire from real environmental news anymore. My point exactly.

After seeing the comments from many of these readers, I noticed a pattern emerging. The people that believed the fake stories tended to be those who are most worried about the environment. In contrast, those who were more discerning and recognized that the stories were bogus tended to be much less worried about environmental doom in general.

Now, I'm not implying that environmental worriers tend to be less critical thinkers. Oh, okay, yes—I am implying that.

The comments from some of these people were considerably funnier than what I could have dreamed up for my bogus news articles. In one story, I "revealed" a recently declassified spy satellite photo that showed islands in the Bermuda Triangle actually levitating above the surface of the ocean. It led to the following comment from someone called BlueDolphin:

> That is absolutely bizarre! I know that there are strange gravitational forces in that area, due to the charged crystals of Atlantis still generating from where they sit on the Ocean floor, I wonder if somehow the crystal vortexes are interfacing with the additional energy anomalies caused by HAARP's activation, and that is the reason for this ...?? most curious....

One lady posted to a new-age forum that she was worried about what these levitating islands might indicate about the Earth's health. In an attempt to console her, an obviously wise and informed Lady Kadjina replied:

> Beneath the waters of the Bermuda Triangle are huge pillars and great crystals as well as the landing fields for many crafts. Jacque Coustou [*sic*] took photographs of all this and many years ago it was shown on your Television. Only one time and then it was banned. You are made from earth and you both are constructed in the same basic way, with

> meridians, acupuncture [*sic*] points, chakra system and both of you are sentient beings. Mother Earth is sentient. All is well. Do not be afraid.

I have to wonder what kind of herbs that lady grows in her garden.

In another article I wrote, fictitious dolphin researchers claimed that a school of dolphins off the Florida coast swimming in a northward direction was evidence that they were fleeing manmade warming of tropical waters. Recognizing the bogus nature of the article, someone posted to a forum his eloquently phrased warning to others that my web site was part of a government disinformation conspiracy:

> This, my friends, is the CIA/Pentagon war against the truth and The People now being waged on the internet just as it is in every other institution and venue by sowing confusion and tripping up well-meaning people who unknowingly spread it like a communicable disease.

It is more than a little weird to be accused of being part of a "CIA/Pentagon war against the truth." One woman who claimed to be a dolphin researcher e-mailed me and asked for additional information about the dolphin observations. When I broke the news to her that the article was fictional, she became very hostile. I'm starting to get a little paranoid that my attempts at humor are making some people hate me.

My point is that many people are not very discerning about what they learn from various media sources. Fortunately, despite the biased media coverage of global warming, I find that many others have still remained pretty skeptical of dire global warming claims.

During my travels, the most frequent opinion I hear from people is that our present global warmth could well be mostly natural rather than manmade. Most people seem to understand that climate has always changed, and will continue to change, with or without any help from humans.

In short, the public has grown distrustful of scientific predictions of gloom and doom. Gee, I wonder why? Could it be because, historically, scientists have always been wrong about these predictions? Hmmm.

"But doesn't science tell us how things work?" you might ask. Well, yes and no. The disciplined practice of scientific investigation will usually give us a better idea of how the natural world works than, say, making something up. (You might have noticed from media reports that scientists are sometimes caught doing this, too.) Unfortunately, a variety of practical problems leads to much less confidence in some scientific conclusions than others. And this brings us to a startling fact that you might not be aware of: *science is not truth*.

While experts remain at odds over the issue of when life begins, most agree it's sometime after work

Chapter 2: Science Isn't Truth

THE WORD "SCIENCE" comes from the Latin *scio*, "to know." So, science is *knowledge*. And as most of us older than thirty can attest, what we know isn't necessarily so. In order to begin to understand why there is so much debate about manmade global warming in the science community, you need to first accept that science doesn't provide us with truth. The practice of scientific investigation involves tools to help us explain how the physical world *might* work. The explanation doesn't have to be true to be useful, just consistent with most of the evidence.

In our technologically driven age, people want to believe that all of life's questions will eventually be answered through science. After all, our lives have been made so much healthier and more enjoyable through the inventions and discoveries that the application of scientific investigation has brought us. But there are some areas of scientific study for which it is particularly difficult, if not impossible, to get hard answers.

When science tries to explain what happened long ago, when no eyewitnesses were available to make measurements, I do not consider that to be a "hard" science. Even though paleoclimatologists try to reconstruct the climates of past centuries or millennia though proxy measurements such as tree rings, there is no way to verify how accurate those interpretations are. A very weak relationship that is found between tree rings and temperature over the last hundred years is extrapolated back 2,000 years, and the result is called "science."

Much more confidence can be placed in actual human observations. For instance, the written records of the Vikings who colonized and farmed Greenland during the Medieval Warm Period are pretty indisputable. Similarly, their gradual migration out of Greenland when colder weather ruined crops, and when icebergs began to appear and threaten safe passage of their boats, are also part of the historical record. They may not be quantifiable in terms of a precise temperature, but then neither are 1,000-year-old tree rings.

Scientific progress requires quantitative measurements that can be verified, testing of alternative hypotheses (possible explanations of how things work), and experimentation. But while science deals with observed facts or measurements, scientific debate usually does not arise over the existence of those measurements. Instead, most of the debate usually centers on differing opinions about what the measurements mean, what they are telling us about the way nature works.

And that part of science is the interesting part. Scientists like to figure out the significance of our observations, and what they are telling us about our world. Unfortunately, not all areas of scientific study are created equal; some sciences are blessed with an abundance of ways to test theories, while other sciences do not have this advantage.

While the interpretations from study of past climates might well be true, there is no way to know for sure. No matter how long and how hard science analyzes a problem, the answers might simply be unknowable. As I said, not all areas of scientific study are created equal.

In the case of global warming, we really don't know how

warm the Earth is relative to past centuries, millennia, or eons. Furthermore, as I will explore in the coming chapters, even though warming is actually occurring today, science still does not have a way to reliably discriminate between manmade warming and natural warming processes. We cannot put the Earth in a laboratory and carry out experiments on it. There is only one global warming experiment, and we are all participating in it right now.

Nevertheless, for reasons ranging from economic to human survival, mankind still needs answers about future levels of warming. Science must do what it can to provide some of these answers as best it can. Scientific uncertainty will always exist, and so policy decisions will have to be made in the face of scientific doubt.

But as is often the case with fields of study that have such strong political, economic, and even religious connotations, our emotions can lead us to overstate the ability of science to provide the answers that we are so desperately seeking. People start to misuse scientific research results as an excuse to facilitate social or political changes that they wanted to see happen anyway. I guess this is just human nature, even for scientists.

The Human Scientist Theory

My wife does not agree with me on this, but I have a theory that scientists are human. Scientists have the need to believe that the research they are doing is important. They have religious, economic, and political biases and opinions—their own worldview. Scientists can get emotional and defensive when their research is challenged. That, in fact, is pretty common.

We scientists can usually be divided into two main camps—male and female. We also have a wide variety of other characteristics in common with regular humans. But in contrast to most humans, who must provide useful goods and services in their jobs in order to earn a living, the government-funded scientist's job is to spend your money. As I will explore in Chapter 6, this tends to make most scientists relatively clueless about basic economics.

Some of them then make stupid pronouncements about what should be done about this or that problem that society is presented with.

There are some subjects that are sure to cause an argument: religion, politics, war, money. Science doesn't seem like it would be one of them. But just as the subject of evolution will cause two people who normally see eye to eye to start arguing, tempers also flare when global warming is brought up in conversation. I have participated in several internet forums that have nothing to do with climate or global warming. As soon as the topic of discussion strays into global warming territory, it is almost guaranteed an argument will ensue. And just like these heated disagreements between humans, scientists also get hot under the collar on the subject of global warming. See? Even more evidence to support my human scientist theory.

But shouldn't scientific inquiry be dispassionate and objective? After all, science doesn't really care what the answer is to a scientific question; it just provides tools for us to try to find the answer. Yes, that's the way it should work, but it seldom does. So, when scientists become emotionally attached to a specific theory, you know that more than science is involved.

Like most other people, scientists don't know as much as they pretend to know. Scientists generally don't like to reveal to the public the uncertainties that are associated with their research. It might make them look less expert. Sometimes it is just too complicated to explain all of the uncertainties. Whatever the reason, claims that scientists make are usually more dramatic and confident than can be defended with the science alone.

Furthermore, since it is only a relatively few scientists who are willing to speak out publicly about global warming, these tend to be the ones that make the more dramatic claims. If they didn't, most reporters wouldn't give them the time of day. And guess which researchers have the most influence on government funding managers and members of congress?

Scientists' personal biases inevitably lead to friction and divisions in the scientific community. As a result, one scientist who researches the effects of warming on hurricanes has accused an older, more famous hurricane researcher of having "brain fossilization." Another climate scientist refused to present a talk after learning that a scientist with whom he disagreed would also be giving a talk in the same conference session. One very famous global warming scientist refused to testify in a congressional hearing when he found out that a global warming skeptic would also be testifying.

Furthermore, every scientist likes to think that the problem that he or she is working on is important to humanity. Who wants to devote their life's work to something that no one cares about?

> *Scientist:* Honey, I'm home!
>
> *Spouse:* Hi, dear. Did you discover anything exciting today?
>
> *Scientist:* Oh, yeah! I found that the tsetse fly actually does a little dance before mating! I can't wait to tell everyone at our next international conference!
>
> *Spouse:* That's nice, dear.

This leads to a tendency for scientists to exaggerate the certainty and importance of their conclusions:

> *Reporter:* So, Dr. Scientist, what are the implications of this finding about the behavior of the tsetse fly?
>
> *Scientist:* Well, by understanding what behaviors lead to mating in the fly, we hope to better understand the original human evolutionary process, how the first male and female humans ended up "getting together," as it were.

The final result is then the news story written by the reporter in the *Daily Rag*:

> *Headline:* "Mating Behavior in First Humans Revealed by Fly's Dance"

And if the behavior of flies is newsworthy, what could be more important than Saving the Earth? Even if a particular scientist's research has not been particularly Earth-shaking, they will typically allow themselves to be prodded by a reporter into overstating their conclusions. As a result, the "truth" of global warming then gets so repeated, mutually supported, and inbred between the media and the global warming pessimists that increasingly bold claims appear in the news.

It has now reached the point where you will hear claims like these: "all reputable scientists agree," "skeptics are like those who would deny the Holocaust or the dangers of smoking," or "skeptics only take their position because they are funded by Big Oil." Some climate scientists would have you believe that manmade global warming is more than just a theory—it is a fact. This is a dead giveaway that those scientists have an emotional attachment to the issue—yet another indication that they are human. While actual thermometer-measured global warming might indeed be considered an observational "fact," *manmade* global warming is far from it.

In their efforts to convince you that manmade global warming is serious, some scientists will even appeal to the public's love of animals—at least the cute ones. When a TV special or movie on global warming suggests that global warming is causing polar bears to drown, our emotions overcome our sensibilities. (I haven't heard yet whether global warming threatens slugs.) Most reporters fail to mention the fact that the total polar bear population has grown dramatically in recent decades. Even Al Gore's movie couldn't find real video of a polar bear threatened by a lack of ice—they had to create a computer animation of a poor bear swimming in an ice-free sea.

This appeal to our emotions is part of what constitutes news today, and for many issues it doesn't matter a whole lot whether the problem we choose to believe in is real or imaginary. But when it comes to a subject as important as global warming, we need to separate our emotional attachment to an idea from what we know (or don't know) based upon the science.

Uncertainty in Science

Nothing has ever been proved for sure in science. Most scientists don't even realize that science itself involves some basic assumptions (postulates) that cannot be proved, only assumed. Yet these postulates are necessary for science to progress. One is that the universe is real, and that humans are capable of discerning its real nature. Another assumption is that nature is "unified," that is, that the physics we measure here and now are the same as the physics operating at other locations and at other times. These are things we assume to be true when carrying out our science, but there is no way to prove them to be true.

A very common trap that scientists tend to fall into is forgetting all about their assumptions. In order to address any problem quantitatively, the scientist must first make simplifying assumptions. If assumptions aren't made, it is usually too difficult to analyze most physical problems. By the time the research is completed and the conclusions are finally made, though, the scientist typically forgets all about his original assumptions. This is probably the biggest single source of the scientist's overconfidence in his conclusions. It is also the first startling discovery I made as a fresh young researcher about other scientists that eventually led to my theory that scientists are human.

This is not to say that uncertain scientific research results are not valuable. While the scientific method is not strictly applicable to every kind of research, it does represent a series of steps that the researcher should take to minimize the chance that he or she will come to the wrong conclusion. Formulate a hypothesis of the way things work. Devise an experiment to test your theory. Make measurements, and analyze the data to see if they support your theory. Our methods of scientific inquiry are pretty good at improving our chances of not falling for logical fallacies or happenstance while trying to discern how nature works.

But scientific inquiry isn't foolproof. Even if the data happen to support your theory, it could be that they support someone else's theory even better. Not even scientific "laws" have been proven to be true. A physical law is simply a theory which scientists have grown tired of trying to disprove. As an example, there used to be a law in nuclear physics called the "Law of Parity" that involved the weak atomic force. It was a law, at least, until some clever researchers disproved it in 1956.

As another example, a 2005 Nobel Prize was awarded to Australians Barry Marshall and Robin Warren for their discovery of the bacterial basis for peptic ulcers. The consensus of medical opinion used to be that stomach ulcers were the result of a stressful lifestyle or too much spicy food. Marshall had the audacity to suggest at a 1983 conference in Brussels, Belgium, that ulcers might instead be caused by a bacterial infection. As Marshall recounts, this was widely considered to be "the most preposterous thing ever heard." It is not easy to overturn scientific "truths," and it took more than two decades before this startling claim led to the highest honor a researcher in any field can receive.

One of the nagging uncertainties that science always has to deal with is that of attributing causes to observed effects. This is true in all scientific fields, especially medical research. Performing science usually entails making numerical measurements of some sort, which are then analyzed for statistical relationships.

It is relatively easy to establish whether a relationship exists or not, but the difficulty comes when we try to explain why it exists. For instance, a researcher might find that, out of a study of 10,000 adult alcoholics, 97 percent of them drank milk as a child. The researcher might then hypothesize that the drinking of milk as a child leads to alcoholism later in life. But as you might suspect, there are alternative explanations as well. It could have been the cookies that were eaten with the milk.

Peptic ulcers are hardly a contentious philosophical, political, or economic issue. They are carried around by millions of people, so we know they are real, that they exist. They can be seen, through optical instrumentation, by the human eye. In contrast, manmade global warming is a mental construct. There is only one possible case of it on the Earth, and the observational evidence for it is obscured by all of the other chaos that the climate system is creating at any given time.

The best way to build confidence in a scientific theory is to test the predictions of that theory against measurements. The trouble with global warming theory is that we cannot test it in the laboratory. What we want to know is how the climate system will respond to increasing levels of greenhouse gases. There is only one experiment going on, and we cannot prove that the warming we have been experiencing has been due to those greenhouse gases or some natural change in the climate system.

The closest thing to a natural climate experiment that we have been able to measure was the 1991 eruption of the Mount Pinatubo volcano in the Philippines. The millions of tons of sulfur it spewed into the stratosphere caused a 2 percent to 4 percent reduction in solar radiation in the Northern Hemisphere. This was followed by one to two years of cooler than normal temperatures. This is viewed by some as providing a quantifiable example of how the climate might respond to more greenhouse gases. But sunlight is the source of energy for the climate system, while greenhouse gases (which we will discuss in more detail in Chapter 3) determine how energy is redistributed in the system.

Yes, there has been globally averaged warming in the last thirty years. Yes, greenhouse gas concentrations in the atmosphere have increased in the same period of time. But this does not prove that drinking milk as a child causes alcoholism later in life.

Certainly a majority of climate scientists would agree that global warming is a potential problem in the coming century. But when you hear the phrase "all reputable scientists now agree," then you can be pretty sure we're not talking about something that has in any way been "proved." Very little global warming research actually results in a conclusion that the evidence supports mankind as the cause of current global warmth, rather than some natural process. Instead, most published research on manmade global warming simply assumes that it exists—not that it doesn't exist. As a result, that research appears to "support" manmade global warming. But that is what the research was funded to study.

A widely publicized study by Naomi Oreskes in 2004 claimed that of 928 abstracts of published research articles dealing with "climate change," none were found that disputed the scientific consensus that recent global warming can be attributed to humans. Aside from the fact that I have a stack of such papers in my office, I would wager that neither did any of those 928 articles demonstrate that our current global warmth is not due to natural causes. Manmade global warming is simply assumed to be true because we have no reliable way of observationally separating natural sources of global warming from human sources.

Maybe the "fact" that the Earth has warmed can be considered to be "truth." *Why* the Earth has warmed, though, is another matter entirely. If you want possible physical explanations for what we observe in nature, go to science. If you want truth, go to church.

Next, I would like to give you a crash course, Weather & Climate 101. Don't worry, there are no tests, and I will keep it as simple as possible. Just bear with me, and by the end you will have a better appreciation for just how complex the climate system is. Then, you can judge for yourself whether science knows enough to claim that "the science is settled" on manmade global warming.

... now I'll pointlessly show the isobar map as usual

Chapter 3: How Weather Works

WHILE MOST BOOKS on global warming try to convince you that this or that scientific study shows evidence for or against manmade global warming, that feels too much like a contest to me. It's as if whoever can list the most published research findings supporting their side wins. But science isn't about winning debates, or taking a vote, or forming a consensus. The climate system is, or is not, sensitive to mankind's greenhouse gas emissions.

So, rather than covering an endless list of specific scientific papers and what they claim to have discovered about climate change, I instead want to equip you with a basic understanding of how weather, and thus climate, works. I want you to appreciate how complex the climate system is, how little we really know about it, and what its most fundamental purpose is: to get rid of excess heat. Finally, I will describe what I believe to be the thermostatic control mechanism that will limit the amount of climate change we will experience from human activities.

By teaching you the basics of how weather operates, I hope to make you informed enough so that you can think about the atmosphere and how it behaves and then make your own judgments. This is better than to ask you blindly to accept some scientist's claim that humanity has only ten years left to do something before we are all doomed.

So to get you going on this little lesson, let's start with weather. Weather is the source of endless fascination for me, and probably for many of you as well. It seems that everyone is interested in the weather, especially those who live in areas that are subject to severe weather threats. And that would include just about everyone. Severe thunderstorms, hurricanes, tornadoes, windstorms, floods, droughts, hail, lightning, snowstorms—all of these make us want to understand how weather operates.

I also have found that nearly everyone has some fundamental beliefs in common about the weather. There are several weather truisms that observe no geographic boundaries. The first is: where you live just happens to be the most difficult place to forecast for in the whole country. Secondly, if you don't like the weather right now, just wait ten minutes. Finally, weather forecasters are incompetent fools. While that last one might well be true, at least weather forecasters can usually explain to you, in learned terms, why they screwed up the latest forecast. We also took classes in college with names like "Effective Weather Lying."

And now, the threat of global warming makes weather even more relevant to our lives, or at least to the lives of our children, grandchildren, and the current crop of politicians and climate scientists. More people than ever are now interested in the weather. This is especially true after they learned that global warming was going to cause even more severe thunderstorms, hurricanes, tornadoes, windstorms, floods, droughts, hail, lightning, snowstorms —all possibly as soon as the day after tomorrow.

I became interested in the weather while in high school because of a buried sewer pipe and some dead sheep. Really. Late in my senior year, we had a "career day" for which we could choose any local governmental office to visit. We would learn first-hand how that office operated and what working there was like. The student assignments to these offices were on a first-come, first-served basis. The National Weather Service Office slot was always the first to go.

I would like to report here that I got that slot, but I didn't. Unfortunately, I've always been a procrastinator. Since I was the very last one in the senior class to sign up, I got what had historically been recognized as the *least* desirable choice: the public health department. While the other students had a day of fun, I traveled around the county that day with a health department employee inspecting a broken sewer line, and then a bunch of dead sheep that a farmer had left in a ditch by the side of a road.

I was soooo jealous of the kid who got to go to the weather office on that day. I think that was the first glimmer of interest in a weather career for me.

What is the difference between weather and climate? Believe it or not, there really isn't any strict definition. It is probably sufficient to say that climate is the average weather for a certain time of year at a certain location. Or, it can be the average temperature for the whole Earth over many years. Either way, climate is just average weather.

For instance, for the last thirty years in Podunk, Michigan, the month of July experienced an average high temperature of 83° Fahrenheit, an average low of 62° Fahrenheit, and 5.6 inches of rainfall. Those are the climatological averages. The average surface temperature of the whole Earth is estimated to be about 57° Fahrenheit. That is the climatological average.

But there is another important distinction between weather and climate, one that helps to explain why forecasters have zero forecast skill beyond about ten days in advance, yet most climate researchers think we will be able to forecast the average climate decades in advance. Weather forecasting is an example of an "*initial value* problem." By measuring the initial state of the atmosphere with weather balloons, airplanes, and satellites, we can, in effect, extrapolate current weather trends into the future by using equations in a computerized weather prediction model. But beyond about ten days, the skill with which we can do this drops to near zero. This ten-day limit has usually been attributed to the "butterfly effect," whereby unmeasured events, even tiny ones like the flap of a butterfly wing, can influence the weather many days later. So, even if you just burp outside, within a few weeks global weather patterns will be totally different than if you had not burped. I'm glad someone named it the butterfly effect instead of the burp effect.

In contrast to this "initial value problem" of weather forecasting, climate forecasting is a "*boundary value* problem." In this kind of forecasting, we examine how small changes in the rules by which the climate system operates can change the average weather. In the case of global warming, mankind is adding carbon dioxide, a known greenhouse gas, to the atmosphere. This will, to some extent, change the way the atmosphere moves heat around in an average sense. Thus, even though climate modelers cannot forecast what the weather will be like on July 4, 2019, they hope to be able to estimate how much warmer the year 2019 will be than, say, 1999. They expect that a small change in the rules by which the climate system operates will translate into an ability to forecast how the average weather (climate) will change.

Now let's examine the most basic processes that determine how our weather operates. We'll use the following simplified illustration, which is appropriate for either middle school students or congressional testimony.

THE SUN WARMS THE EARTH

The starting point is the energy source for our weather: the sun. The accompanying illustration shows that sunlight gets absorbed by the Earth, and this causes weather stuff to happen in the atmosphere (more about that later). Take a few seconds to look at the basic processes in this illustration. Go ahead ... I'll wait.

The one process that you might not recognize in this illustration is infrared radiation. As we shall see, this just happens to be the one process we are most interested in when it comes to global warming.

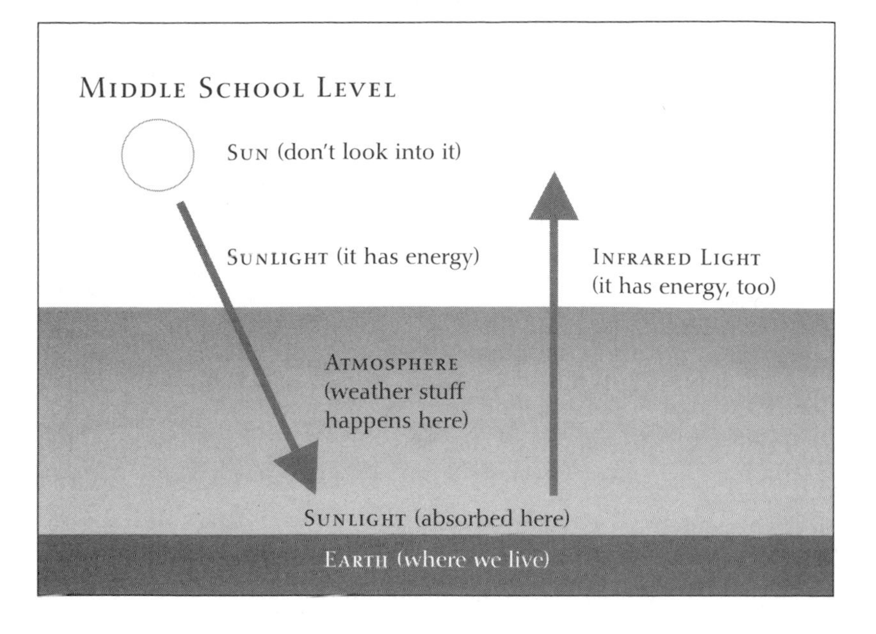

Infrared Light Cools the Earth

In our illustration, sunlight is warming the Earth, and by itself that sunlight would cause the Earth to get continuously hotter and hotter if there wasn't some way for the Earth to also get rid of heat. The way the Earth does this is almost entirely through infrared (or "heat") radiation, which is being continuously emitted by the Earth to outer space.

Even though the habitability of the Earth depends upon it, infrared light is not very well understood by most people. This is probably because it is invisible. But we can feel it. You are no doubt familiar with the infrared radiation you feel at a distance from a fire, or a red-hot stove element. You notice these sources of infrared light because our skin (unlike your eyes) is sensitive to it, especially if the source is very hot.

In fact, *everything* absorbs and emits infrared radiation. The hotter something is the more infrared energy it emits. Even you, sitting there reading this book, are losing infrared radiation to your surroundings. The infrared energy you radiate helps to cool you off in response to your metabolism's continuous generation of heat. You are invisibly glowing. You might have seen simulations of this effect in the hit TV show "24," or in movies, where satellites (magically) see through buildings and sense the infrared heat being radiated by people. Similar technology is used by the military.

So what is the role of infrared energy in weather? In order for the average temperature of the Earth to remain relatively constant, all of the absorbed sunlight must be balanced by an equal amount of infrared energy escaping from the Earth back to outer space. This concept is called "radiative energy balance," and it is central to global warming theory. The Earth's temperature remains fairly constant because the amounts of radiant energy entering and leaving the Earth system are about equal. Any imbalance between the solar input and the infrared output will cause either a warming tendency or a cooling tendency. The bottom line is this: a constant temperature requires that energy in = energy out.

Let's return to the example of your body losing infrared energy to your surroundings. Since your surroundings are also losing energy (in proportion to their temperature), your body is also *absorbing* infrared light at the same time it is *emitting* it. But since your body is usually warmer than your surroundings, you send out greater amounts of infrared heat than your surroundings send back to you. In this case, the infrared energy leaving your body is greater than that being absorbed by your body, and so you are "radiatively cooling."

Now replace your body with the whole Earth (figuratively speaking, of course). The Earth is continuously emitting infrared radiation to outer space, day and night. The infrared energy emitted by outer space toward Earth is almost zero, and can be ignored. The amount of infrared radiation emitted, averaged over the whole Earth over many years, is believed to be very close to the amount of sunlight that was absorbed over the same period of time. Again, radiative energy balance, and so a relatively constant temperature. The magnitude of the solar heating and infrared cooling, averaged over the whole Earth, is estimated to run about 235 watts per square meter (which is about 22 watts per square foot).

As an everyday example of the role of infrared radiation, we all have the experience of the air cooling off after the sun goes down. Strictly speaking, it doesn't cool off at night because there is no longer any sunlight coming in. It cools off because the infrared cooling (energy out) exceeds the solar heating (energy in), which of course is zero at night.

Then during the daytime, even though the Earth is continuously losing infrared heat to outer space, the amount of energy being absorbed from sunlight is greater than that being lost. In this case, the energy in exceeds the energy out, and so everything warms up.

Since our eyes are not sensitive to infrared light, it takes some mental practice to get used to the idea that the Earth is continuously emitting infrared energy to outer space. For some reason, we humans have a difficult time believing something exists when we can't see it. There is a simple experiment you can do to actually feel the Earth cool. On a clear cool evening after the sun has set, stand on a driveway, parking lot, or even a grassy area, that had been heated by the sun during the day. Hold your hand out horizontally, palm facing down. Then turn your hand over, so that your palm is facing up. Keep flipping your hand over, palm up, then palm down. As you do this, you will be able to sense the different amount of infrared energy coming up from the warm ground versus coming down from the cold sky. *Voilà*, you are now an infrared radiometer.

Nighttime infrared cooling of surfaces exposed to the cold sky explains why dew forms on cars, grass, and other surfaces. Objects placed under a tree will stay warmer at night because those objects are being heated by infrared radiation from the tree, rather than losing so much infrared energy to the relatively cold sky.

Another aspect of infrared radiation that makes it more difficult to conceptualize than sunlight is the fact that *everything* is continuously emitting and absorbing infrared heat. Whereas only the sun can emit sunlight, infrared radiation is being continuously emitted (and absorbed) by everything—buildings, trees, grass, air, clouds, etc.

Because the upward and downward flows of infrared radiation in the atmosphere are so complex, our intuition fails us when we try to figure out how they affect the temperature structure of the atmosphere. Instead, we have to run what is called a radiative transfer model in a computer that contains the relevant physics. Don't try this at home ... leave it to trained professionals.

This is why global warming theory is a little difficult to understand for layman and expert alike. Global warming theory involves how infrared energy is redistributed within, and lost by, the surface and atmosphere, and those processes are totally invisible.

While we will address global warming theory in more detail in the next chapter, at this point I am just trying to get you used to the idea that solar heating and infrared cooling are what "drive" our weather. And those infrared processes in the atmosphere are what lead to the Earth's natural "greenhouse effect."

THE NATURAL GREENHOUSE EFFECT AND WEATHER

Greenhouse gases are those atmospheric gases that strongly absorb and emit infrared energy. If your eyes were sensitive to the infrared wavelengths of light that greenhouse gases absorb and emit, you would not be able to see very far. Looking around, you would see an infrared "fog" everywhere, as if you were in a cloud. The major greenhouse gases in the Earth's atmosphere are water vapor (which accounts for about 70 percent to 90 percent of the Earth's natural greenhouse effect), carbon dioxide, and methane. Additionally, clouds also have a large greenhouse effect, but clouds are not a gas. They are made up of tiny liquid water droplets or ice crystals.

The most important thing to remember about greenhouse gases in the atmosphere is that they act like a blanket, making the lower atmosphere warmer, and the upper atmosphere cooler, than those layers would otherwise be without the greenhouse gases. This is somewhat analogous to a blanket covering your body. The blanket keeps you warmer and the air on the outside of the blanket cooler, than if the blanket was not there. But whereas a blanket primarily works by preventing the movement of the air that is heated by your body, a greenhouse gas is a "radiative blanket" that keeps the lower atmosphere from cooling too rapidly.

Even most meteorologists don't realize this, but the existence of our weather depends upon the greenhouse effect. The dotted line in the following diagram shows how the temperature of the troposphere (the lowest layer of the atmosphere, where our weather occurs) would change with height if there was no weather. The average surface temperature of the Earth would be around 140° Fahrenheit, and the altitudes at which jets fly would be so cold that their fuel would gel. Since we can't actually prevent the atmosphere from producing weather, this is a theoretical calculation made from a radiative transfer model.

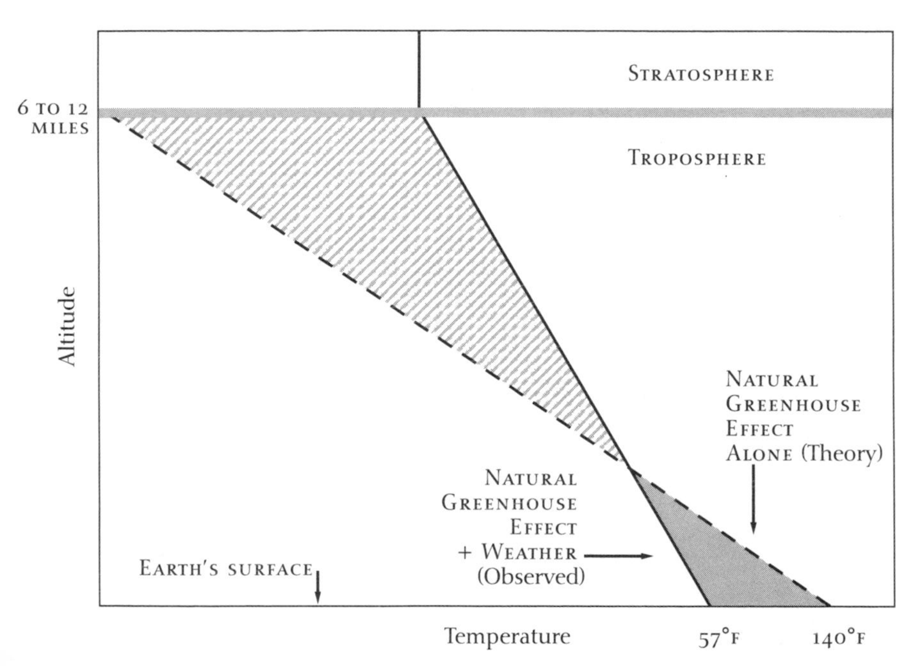

The Earth's natural greenhouse effect "wants" to make the Earth's surface unbearably hot, but the cooling effects of weather prevent most of that warming from occurring.

This 140° Fahrenheit greenhouse temperature is, I believe, the best starting place to explain what drives our weather. The combination of solar and infrared radiation together "tries" to make the Earth's surface extremely hot. But long before that temperature state is reached, the atmosphere becomes "convectively unstable," which just means that warm air starts to rise and cooler air starts to sink. The real atmosphere overturns continuously, transporting excess heat from the surface to high in the atmosphere. All of the processes associated with this overturning are part of what we call weather.

The shaded region in the lower part of the diagram represents how much temperature decrease is caused in the lower troposphere by weather processes. The hatched region in the upper troposphere represents the temperature increase that results from weather processes transporting that heat up from below. The most dramatic examples of this transfer of heat from the lower troposphere to the upper troposphere are thunderstorms and hurricanes.

Why am I emphasizing the fact that weather cools the surface to a temperature well below what sunlight and the greenhouse effect are trying to make it? Because, while many people have heard that "the greenhouse effect makes the Earth habitably warm," virtually no one has heard that "weather makes the Earth habitably cool." Quantitatively, the cooling effects of weather are actually stronger than the greenhouse warming effect. So, why is it that we never hear about that in discussions of global warming? Hmmm?

If you have made it this far, then congratulations. (Of course, if you haven't made it this far, you aren't even reading this.) You are now well on your way to being a weather expert—or as I like to say, a weather weenie. I am promoting you to the high school level of understanding.

Heat Removal from the Earth's Surface

We have addressed how incoming sunlight is the source of energy for our weather, and how infrared light provides a way for the Earth to lose excess energy to outer space. Next we learned that the combination of solar heating and infrared energy transfers are continuously trying to make the Earth's surface unbearably hot and the upper atmosphere unbelievably cold. What happens in response to all of this "radiative forcing" is where all of the interesting stuff that we call "weather" happens.

Even most weather and climate people don't really think about the ultimate purpose of what we call weather: *to move heat from where there is more, to where there is less.* Every gust of wind that blows, every cloud that forms, every drop of rain that falls, all happen as part of processes which continuously move excess heat from either the surface to higher in the atmosphere, or from low latitudes (tropical regions) to high latitudes (polar regions).

These flows of heat are a demonstration of one of the most basic laws in science—the Second Law of Thermodynamics—which in simple terms just states that energy tends to flow from where there is more to where there is less.

Now we are ready for the high school version of our weather illustration. It shows an idealized thunderstorm extending through the full depth of the troposphere. While there are a number of pathways that energy can follow in the process of creating our weather, here I will describe the dominant one. We begin at the Earth's surface, since that is where most of the sunlight entering the atmosphere is absorbed.

In our illustration, the land surface and upper ocean are warmed by the sun. But the resulting temperature increase is not the same everywhere. The tropics receive more sunlight, and so are warmer, than the polar regions. Clear regions receive more sunlight, and so get warmer, than cloudy regions. The land warms up faster than the oceans. The point is that the temperature of the Earth's surface is pretty uneven.

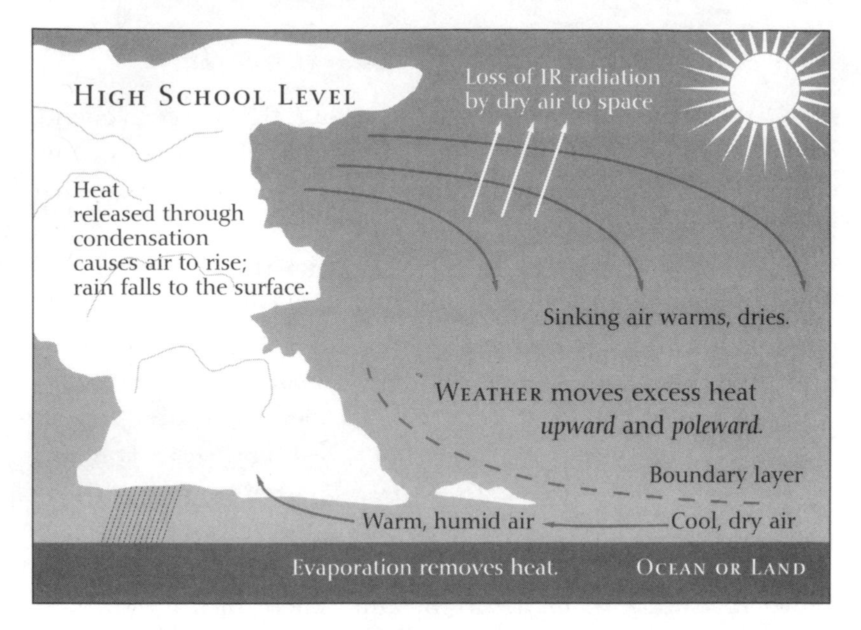

Ultimately, it is the difference in temperature between one region and another that cause air currents (wind) to blow across the surface of the Earth, which then pick up heat from the surface and move it someplace else. The heat transferred from the surface to the overlying air is either *sensible* (an increase in air temperature) or *latent* (water evaporated from the surface, adding water vapor that contains the latent heat of vaporization to the air). Now stick with me here, because this is really important.

Latent heat loss by the Earth's surface through evaporation is the dominant mechanism for cooling it. "Latent" refers to the fact that the heat energy added to the air does not increase its temperature, but is instead used to change the water from its liquid form to its vapor form. The process of evaporation requires energy, and explains why a breeze blowing on your wet skin feels so cold. The water is stealing heat from your body so it can turn into water vapor.

At least 90 percent of the heat lost by lakes and oceans is through the energy required just to evaporate water from the surface. For land surfaces, much of the evaporation occurs from water cycled through plants as part of the growing process, which is called evapotranspiration.

The astute reader at this point might be a little confused about the role of water vapor. I previously had described how water vapor has a strong greenhouse warming effect on the lower atmosphere. Now I have just told you that the evaporation of water is the dominant cooling mechanism for the surface. So, which is it? Does water vapor cool the surface, or warm the surface?

The answer is—it does both. When the surface water is evaporated to form vapor, it removes heat from the surface. After that, the vapor then helps warm the surface through the greenhouse effect. Of course, both of these effects are happening at the same time, continuously. Water is a miraculous substance, performing a wide variety of functions in weather and climate. Even if surface water is polluted, once it evaporates it is then once again pure, ready to perform its assigned tasks all over again.

If neither surface water nor vegetation is present, then all of the sun's energy is turned into sensible heat (a temperature rise). Without water to absorb some of that heat through its conversion to vapor, all of the energy goes into raising the temperature of the air instead. This is what causes the "urban heat island effect" in cities, and it also explains why temperatures in the desert get so hot. There is very little water to absorb the heat through evaporation.

To help counter the urban heat island effect, some cities are encouraging the planting of vegetation on the roofs of buildings and elsewhere. This helps convert more of the absorbed sunlight into latent heat (stored in the vapor) rather than sensible heat (stored as temperature).

As a side note, deserts are not hot because the sand is so bright. The brightness, in fact, keeps the desert air cooler than if the desert sand was black, by reflecting more sunlight back to outer space. If we painted everything in our cities white, they would not absorb so much sunlight, and would stay much cooler. The brightness, however, would probably be unbearable.

HEAT TRANSPORTED UPWARD IN THE ATMOSPHERE

All of the heat being lost by the surface and accumulating in the lowest layers of the atmosphere results in parcels of warm air rising, and cool air sinking. If the warmest air parcels rise far enough, their temperature becomes too cold to keep all of the water vapor in its vapor form. Now at 100 percent relative humidity, some of the vapor starts to condense (convert back to its liquid form) as tiny cloud droplets. It is at this precise moment, when a cloud is formed, that the latent heat that was lost by the Earth's surface during evaporation is released, and the air is warmed.

This warming from condensational heating then causes the cloudy air parcels to continue their ascent even higher. You might have felt this rising air while flying in and out of clouds in an airplane. The bumpy ride is due to the latent heat that is being released as some of the water vapor turns into cloud water. If you look out the window at the airplane's wing, you can get some idea of how much heat has been released by how well you can see the tip of the wing when you fly through the cloud. In some cases where a lot of water vapor has condensed into cloud water, the cloud will be so thick that you won't be able to see the wing.

If the warm, cloudy, rising air contains enough water vapor and ascends high enough, the cloud water droplets grow and combine to form raindrops. If the air is sufficiently cold, snowflakes are formed. At altitudes near the freezing point (32° Fahrenheit), rain and snow can occur together. You might be surprised to learn that, even on a hot summer day, the upper parts of thunderstorms are actually mini-snowstorms.

Some of the precipitation then usually makes its way to the ground. Any water that doesn't reach the ground as precipitation eventually re-evaporates to humidify the air once again. Since this re-evaporation absorbs as much latent heat as was released when the vapor condensed into cloud, there ends up being no net warming of the atmosphere. It is only when precipitation actually reaches the ground that a net warming of the atmosphere is realized.

Therefore, every drop of rain, and every flake of snow, that reaches the ground represents absorbed solar energy that has been transferred from the surface to the upper atmosphere.

It is one of the curiosities of weather systems that, even though all of the moist air ascending in cloud systems releases huge amounts of heat, those rising air currents end up being at about the same temperature as the air surrounding them. This is because the warm, cloudy air parcels rise and cool by expansion so that they remain near the same temperature as the surrounding air. As long as they are warmer than their environment, they will rise. When the rising air parcels cool to the point that they are the same temperature as their environment, they stop rising.

So, if rising warm, cloudy air is always cooled by its ascent, how does the upper atmosphere experience a net temperature increase from all of this upward heat transport? The answer is, *in response to all of the rising air, an equal amount of air somewhere else is being forced to sink*. It is in those sinking regions where the greatest amount of air experiences a temperature increase.

This sinking almost always occurs over much larger areas than the rising air within cloud systems. As a result, a small area of rapidly warming and rising cloudy air can cause very slow sinking and weak warming of air over a large area. The most extreme example of concentrated sinking and warming over a relatively small region is in the eye of a hurricane.

Almost without exception, this sinking air is cloud-free, and has low humidity since much of its water vapor has been wrung out as precipitation. Even most meteorologists and climate experts don't realize that when we experience a sunny day with a clear blue sky, it is because precipitation systems somewhere else are forcing the air overhead to sink.

As is shown in the previous illustration, we have now followed the heat from the surface, where the sunlight was originally absorbed, to the air flowing over the surface and picking up some of that heat, to the cloudy ascending air currents where precipitation is formed, to the descending air currents where the actual temperature increase of the middle and upper troposphere takes place. This is the dominant pathway by which heat is transported from the Earth's surface to the upper troposphere, thereby cooling the surface and lower troposphere. There is now one more step in the heat transfer process.

THE HEAT IS LOST TO OUTER SPACE

As the last part of this process, the clear, warm, dry sinking air cools by emitting infrared radiation to outer space, thus completing the cycle of energy into, though, and back out of the atmosphere.

I have neglected some of the weaker pathways by which some of the energy follows. For instance, a small portion of the heat that builds up at the Earth's surface is lost directly to outer space in the form of infrared radiation, thereby bypassing the sequence of events I just described. Proof of this kind of heat loss is in the infrared satellite imagery you see on the TV or the internet. Those satellite sensors are design to sense infrared energy at wavelengths where the atmosphere is transparent, and so they can see infrared radiation coming directly from the ground.

ATMOSPHERIC CIRCULATION SYSTEMS

Note that the processes of heat transfer we have just described constitute an entire atmospheric circulation system. Air picks up heat from the Earth's surface, releases it as it rises in precipitation systems, then flows away from the precipitation systems and slowly sinks and radiatively cools before it once again reaches the surface to start the whole process all over again.

Even though our high school-level illustration shows what appears to be a warm season circulation system over a rather limited region, in reality some tropical circulations can extend for thousands of miles. In the wintertime, outside of the tropics, the ascending moist air flows in a slantwise fashion, rather than straight up, covering large areas and traveling hundreds or thousands of miles to get from the surface to the upper troposphere. These flows occur in association with low pressure areas called extratropical cyclones whose main function is to carry excess heat from the tropics to the higher latitudes.

These are the lows that produce large precipitation shields, which then ruin your October weekend. If you experience sunshine on one day, and rain on the next, this is most likely due to the ascending and descending branches of a single low pressure/high pressure circulation system moving across your area.

To further complicate things, the turning of the Earth causes the air in most large circulation systems to flow *around* high and low pressure areas, rather than to travel directly from high pressure to low pressure. This is called the Coriolis effect. In the Northern Hemisphere, air flows in a counterclockwise direction around low-pressure areas. In the Southern Hemisphere, it flows clockwise. Very close to the equator, air simply flows from high to low pressure.

(And, no, the Coriolis effect does not cause the water draining down your sink to spin in one direction. The sink is too small, has too many irregularities in its shape, and the water flow happens too rapidly for it to "feel" the turning of the Earth underneath it. But experiments with a large, perfectly cylindrical water tank with a very small drain hole in the exact center have shown that, over a period of hours, the water in the tank does indeed spin in only one direction.)

All of these circulation systems, whether in the tropics or high latitudes, are continuously occurring on a global basis. The atmosphere never stops overturning and flowing from one place to another. It is constantly removing heat from the surface and depositing it high in the atmosphere, and carrying it from tropical latitudes where more sunlight is absorbed, toward the poles where less sunlight is absorbed. And remember, all of these weather elements are fulfilling one ultimate purpose: to move heat from where there is more, to where there is less.

Now that you understand the basic processes involved in the operation of weather, we are now ready to address global warming. If you don't understand these basics ... well, just pretend that you do.

Chapter 4: How Global Warming (Allegedly) Works

As previously described, a greenhouse gas strongly absorbs and emits infrared radiation. The dominant greenhouse gases in the atmosphere are water vapor, carbon dioxide, and methane. Together, these greenhouse gases act like a radiative blanket, causing the lower atmosphere to be warmer, and the upper atmosphere cooler, than if they were not there.

Carbon Dioxide Concentrations Are Increasing

The major concern in global warming is that mankind's burning of fossil fuels is slowly increasing the carbon dioxide content of the atmosphere. Those who fret over such things usually put the increase in the most dramatic terms possible, for instance total global emissions are now running about 30 billion tons per year. Notice that they don't tell you is how that compares to the total weight of the atmosphere: 5 quadrillion tons.

But since all those "-illion" words (million, billion, trillion, quadrillion) sound the same, lets look at our carbon dioxide emissions in another way. The accompanying graph shows the upward trend in the atmospheric carbon dioxide concentration since 1958 at the Mauna Loa Observatory in Hawaii. Mauna Loa was chosen as a monitoring site because it is relatively isolated from any major urban areas, which tend to have elevated concentrations of carbon dioxide. Other carbon dioxide monitoring stations around the world show basically the same upward trends.

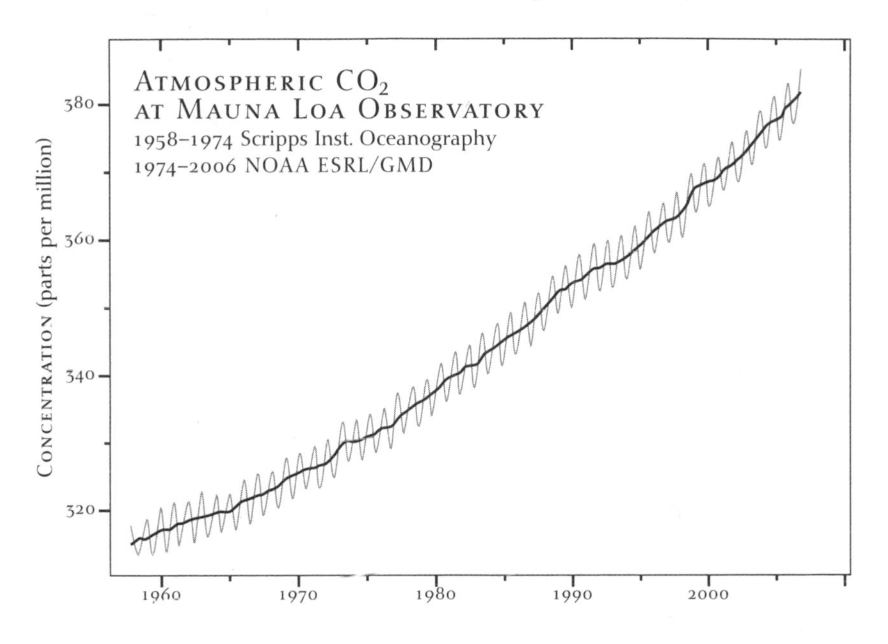

While the rise in the atmospheric CO_2 concentration in this figure looks dramatic, note that the units of concentration are parts per million (ppm). The current concentration of about 380 ppm means that for every million molecules of air, 380 of them are carbon dioxide. Or alternatively, for every 100,000 molecules of air, 38 of them are carbon dioxide. This small fraction reveals why carbon dioxide is called one of the atmosphere's "trace gases." There simply isn't very much of it.

At the rate of rise shown in this graph, mankind adds only 1 molecule of CO_2 to every 100,000 molecules of air every five years or so. This, then, is what is supposedly going to cause a global warming catastrophe. Really—a whole bunch of scientists say so.

Since it is a greenhouse gas, the extra carbon dioxide is believed to be causing a surface "warming tendency" because it makes the Earth's natural greenhouse effect a little stronger—the "radiative blanket" is slightly denser. As a result of this slightly denser blanket, not quite as much infrared energy is being allowed to escape to outer space. This means that the Earth's radiative balance between absorbed solar energy and emitted infrared energy has been disrupted, and more energy is now coming in than is going out. Global warming theory says that the Earth's atmosphere must heat up as a result. Since a warmer atmosphere will emit more infrared energy to outer space, the Earth's radiative energy balance will be restored only when the temperature warms up sufficiently. This is the global warming party line.

Note that careful climate scientists will say that the extra carbon dioxide causes a warming "tendency." But as the philosopher David Stove pointed out in his book *Darwinian Fairytales*, the word "tendency" is ambiguous. Climate scientists use the word because, as in all complex systems like the Earth's climate, a single change can be expected to cause other responses, too. More often than not, those responses act to dampen the original (warming) tendency, counteracting it with other, offsetting tendencies.

But while many scientists and the mainstream media have made global warming sound so very serious, you might be surprised by how small the infrared trapping effect of the extra carbon dioxide really is. When we reach a doubling of the pre-industrial atmospheric carbon dioxide concentration, probably late in this century, we will have enhanced the Earth's natural greenhouse effect by about 1 percent.

"But" the climate alarmist might protest "we are upsetting the Earth's delicate energy balance!" Well, as I will address later, I don't believe that the presumed energy balance of the Earth is really all that delicate. And when scientists tell you that the extra carbon dioxide is causing a surface "warming tendency," it is also very misleading. As described in the previous chapter, the combination of solar heating and the atmosphere's natural greenhouse effect is *always* causing a strong warming "tendency" at the surface and in the lower troposphere—but weather processes keep most of the resulting temperature rise from ever occurring! So, the addition of extra carbon dioxide does not really cause a "warming tendency"—it merely enhances the large pre-existing warming tendency by a tiny amount.

Now let's get beyond all of this hand waving, and talk about real global warming numbers. It has been calculated (again, theoretically) that if there were *no other changes* in the atmosphere, a doubling of the carbon dioxide content would result in only 1° Fahrenheit surface warming. This "direct" warming effect is relatively small, and would likely be easily adapted to by both humanity and by nature.

As we shall see, though, most global warming estimates are much larger than this: from 4 to 10° Fahrenheit warming by 2100. This is because it is widely believed that weather processes will amplify the relatively small amount of CO_2-only warming.

So let's review the basic concepts involved in global warming. Carbon dioxide is a greenhouse gas, which means it tends to warm the lower atmosphere. Mankind is putting more and more if it into the atmosphere from combustion of fossil fuels, at a rate of 1 molecule of CO_2 per 100,000 molecules of air every five years. This causes a slight change in the (alleged) radiative energy balance at the top of the atmosphere. This imbalance, so the theory goes, causes the atmosphere to warm up until the out-flowing infrared radiation that cools the Earth once again balances the absorbed sunlight that warms the Earth. This is the basic explanation of how manmade global warming works.

One might wonder, how do we know that this small radiation imbalance at the top of the atmosphere from the extra carbon dioxide really exists? Well, we don't really know. It is (sigh), once again, a theoretical calculation.

A series of NASA satellites have been flown in recent decades to measure the amount of sunlight being absorbed by the Earth, and the amount of infrared energy being lost by the Earth to outer space. But the expected imbalance between them is still very small as of this writing—a little less than 1 watt out of the 235 watt average. The satellite instruments are not quite accurate enough to measure such a small imbalance with confidence. It would be like trying to see the difference in room brightness when your ceiling is covered by 234-watt light bulbs spaced three feet apart, instead of 235-watt light bulbs.

Another measurement difficulty is that the satellites cannot measure the whole Earth at once. Even though one half of the Earth is absorbing sunlight, and the entire Earth is emitting infrared radiation, the satellite can only measure one small area at a time as it orbits over different geographic regions. For any given place and time, the imbalance between incoming sunlight and outgoing infrared energy is usually very large: many tens, if not 100 watts per square meter. Cloudy weather, clear weather, tropical locations, and polar regions all have characteristically large imbalances between incoming sunlight and outgoing infrared light. So the satellites measure many, very large imbalances at different locations all over the Earth, and the average of all these large numbers together is expected to approach zero (or the very small one-watt imbalance) over a sufficiently long period of time.

The radiative balance concept for the Earth maintaining a constant temperature is really just a theory, based upon some basic physics and some assumptions about the Earth as a whole. For all we know, it could be that there is an average radiation imbalance of a couple of watts per square meter that has existed for hundreds of years which is being continually offset by, say, a small change in the circulation of the oceans. The oceans are capable of absorbing or releasing vast amounts of heat through internally generated circulation changes over very long periods of time.

About all we know for sure about Earth's radiative energy balance is that, for any given place and time, it doesn't exist. All layers of the atmosphere, and all regions of the Earth, are typically very far out of radiative balance. But this isn't a bad thing, because these imbalances are an integral part of what drives our weather.

Like the theoretically-computed one-watt global radiative imbalance, the 1° Fahrenheit direct warming from a doubling of CO_2 is also a theoretical calculation. We use radiative transfer theory together with laboratory measurements of how much infrared radiation is absorbed by carbon dioxide to compute how much more trapping of infrared radiation occurs from increasing carbon dioxide, and thus how much surface warming that extra energy would cause.

This theoretical calculation, however, assumes that the atmosphere does not change in any other way in response to the warming. In reality, though, we know that the atmosphere will react to this warming tendency, just as it reacts to any change that causes a warming or cooling tendency. After all, the primary job of the atmosphere is to move heat from where there is more to where there is less, and if anything happens to affect this, the atmosphere can be expected to respond in some way.

The big question is, will it respond in ways that amplify the small amount of direct warming from the extra carbon dioxide, or dampen it? Those responses are called *feedbacks*: changes in clouds, precipitation, and anything else that alters the direct warming effect of the extra carbon dioxide in such a way that makes the warming either stronger or weaker.

There is very little scientific disagreement over the fact that the extra carbon dioxide mankind is emitting is causing a slight enhancement of the Earth's natural greenhouse effect. What is disputed is how the atmosphere will respond in terms of feedbacks.

The Climate Response to Extra Greenhouse Gases

We have now arrived at the crux of the debate over future global warming. How will the atmosphere respond to mankind's 1 percent enhancement (late in this century) of the natural greenhouse effect? In climate models, the responses are usually put in terms of how they affect surface temperature. While these responses are called "feedbacks," I will try to minimize my use of this term, for a couple of reasons. First, it is somewhat confusing that "positive feedback" is bad, and "negative feedback" is good. Positive feedback is when some change in the climate system amplifies a surface warming tendency, and negative feedback reduces a surface warming tendency.

Secondly, "feedback" puts most of the emphasis on surface temperature changes, which feedbacks are always referenced to. But there are all sorts of possible changes in the climate system, and so it is somewhat biased to presume that surface temperature is the main element of climate change. It is better to think of "global warming" in terms of how energy flows throughout the climate system change, rather than just changes in surface temperature. This is why some people prefer to use the more general term "climate change" rather than "global warming."

In most computerized climate models, the infrared trapping due to the extra carbon dioxide causes substantial changes in the modeled climate system: humidifying of the atmosphere, increases in evaporation and precipitation rates, and a net average surface warming of anywhere from 4 to 10° Fahrenheit. Obviously, 10° Fahrenheit of warming by late in this century would concern just about anyone. But can any of these warming estimates be believed?

In science we usually test theories, such as those encapsulated in climate models, by testing how well they predict other, similar events. Unfortunately, there is no directly observable analog in the real climate system with which we can test greenhouse warming theory. As a result, about all modelers can do is to get the models to behave fairly realistically for the average climate we see today, and then hope that they will behave realistically when the radiative effect of more carbon dioxide is added to them.

The trouble is that there are many decisions that must be made in constructing a climate model, many of which are either arbitrary or very uncertain. We don't know which of these are critical to global warming projections and which are not.

The whole problem is further complicated by the fact that the climate system is an example of a nonlinear dynamical system. Such systems are notoriously difficult to model. I once talked with a retired mechanical engineer who had spent his career building computer models of complex mechanical systems that were to be designed and manufactured. He related how engineers could successfully model the individual subsystems that made up the whole machine. But when all of the modeled subsystems were put together, the computer failed to predict how the system as a whole would behave.

It is safe to say that the climate system is considerably more complex than any manmade machine. Even if we become fairly successful at modeling how all of the individual subsystems (clouds, water vapor, precipitation, etc.) operate individually, what assurance do we have that they will behave correctly when they all are pieced together in a computer climate model? And even if that is successful, will they behave realistically when we artificially enhance the model's greenhouse effect by 1 percent?

In the last chapter I described how, at the most fundamental level, the atmosphere and the oceans are constantly adjusting to non-uniform heating by transporting heat from where there is more to where there is less. Heat gets moved from the surface to high in the atmosphere, and from the tropics to higher latitudes. Most of the atmospheric heat transport is accomplished by the circulation systems associated with a wide variety of precipitation systems and high and low pressure areas. Within the ocean this heat transport is accomplished by shallow, regional circulation systems such as the Gulf Stream and Japan's Kurishio Current, the large-scale wind-driven ocean circulation, and deep-ocean flows like the thermohaline circulation which is driven by regional differences in water density due to salt content.

Climate models do a fair job of mimicking some of these average features of the climate system, but some large discrepancies exist as well. For instance, historically there have been huge variations between different models' estimates of how much heat the oceans transfer from the tropics to the high latitudes, even though the models produce similar global temperature distributions. In other words, they can get what appear to be the right temperatures, but for the wrong reasons.

Even if all the models agreed with each other regarding the *average* behavior of the climate system, this is not what we are really interested in when predicting global warming. We instead need to know how the system will respond to small changes in the "rules" (boundary conditions), such as the slight increase in the greenhouse effect from increasing levels of manmade greenhouse gases. Understanding the sensitivity of the climate system to a small poke in the side is a more difficult task than merely being able to describe how the climate system, on average, works. It requires quantitative knowledge of how all the different processes interact with each other and what controls them.

Let's look at some of the changes that climate models predict will happen as a result of increasing carbon dioxide, and see how well we understand those processes. It must be remembered that, while I will address them individually, all of these processes in the real atmosphere are interacting with each other.

WATER VAPOR INCREASES

The increase in water vapor expected with global warming has more confidence among climate modelers than any other change. The conventional wisdom is that the warming tendency at the Earth's surface caused by the extra carbon dioxide will result in faster evaporation of water, and thus a humidifying of the atmosphere. Since water vapor is the atmosphere's dominant greenhouse gas, this would in turn amplify the warming. All of the dozen or so major climate models that are used to predict global warming behave in this way.

There are, however, a couple of reasons to question the strength of this amplification of the warming by water vapor.

First, the greenhouse effect due to water vapor is not controlled by the surface evaporation, but by precipitation systems. The amount of water vapor in the middle and upper troposphere

—the atmospheric layers that contribute most to the natural greenhouse effect—is controlled by complex processes in precipitation systems. Even though surface evaporation tries to fill the atmosphere up with water vapor, precipitation does not allow that to happen. At typical evaporation rates, it would take only a week or so for the atmosphere to approach saturation (100 percent relative humidity).

Instead, near-surface relative humidities average closer to 70 percent, and in the subtropical high pressure zones at an altitude of a couple of miles, relative humidities can be astonishingly dry: 5 percent or less. Precipitation systems, in effect, limit the natural greenhouse effect—most likely, in proportion to how much sunlight is available.

Since the water vapor content of the atmosphere, and thus the natural greenhouse effect, are under the control of precipitation processes in clouds, we cannot really know how much the atmospheric water vapor content will rise with the CO_2-induced warming tendency without also knowing how precipitation systems will change with warming in their efficiency at removing water vapor from the atmosphere.

And it just so happens that the controls on precipitation efficiency are probably the least understood of any atmospheric process. And that which we do not understand, we tend to downplay in importance. We can't include a process in a climate model that we cannot quantify.

Research published in 1994 by Renno, Emanuel, and Stone in the *Journal of Geophysical Research* demonstrated that, by simply increasing the efficiency of precipitation systems, a cooler climate with *less* precipitation is produced. While that study examined an unrealistically large change in efficiency, so was the resulting temperature change: over 13° Fahrenheit! Another study found that by simply changing how big precipitation droplets are, a different climate state resulted. Studies of this type are, by themselves, sufficient to cast doubt on global warming predictions produced by climate models that have critical processes missing.

No doubt there is a wide variety of changes in precipitation system characteristics that would, at least theoretically, change the average climate. Yet, we don't see these climate changes happen in reality. It is much easier to get a model to behave unrealistically than it is to get it to behave realistically. This should tell us something about putting too much confidence in models when there is still so much we do not understand.

There is anecdotal evidence that warm tropical systems are more efficient than their cooler, high latitude cousins. Hurricanes are believed to be the most efficient weather systems at removing water vapor from the atmosphere. At least conceptually, this tendency for tropical rain systems to be more efficient than high latitude systems suggests that precipitation efficiency changes as a result of global warming could constitute a significant climate stabilization mechanism.

And yet, despite this lack of understanding of precipitation, the consensus of opinion among global warming experts continues to be that positive water vapor feedback is well understood. This appears to be one of those cases where scientists claim to understand more than they really do. While it is possible that they are correct, at this point their position seems to be based as much on faith as it is on science.

CLOUD CHANGES

If water vapor increases are supposed to be the best understood climate response associated with global warming, cloud changes are widely agreed to be the least understood. Cloud formation, maintenance, and dissipation are the result of a succession of complex, interacting processes. The treatment of clouds in climate models is necessarily crude because we don't know enough about what controls them, and even if we did, the processes are so complex that it would require much faster computers than we currently have to include those complexities.

Even though modelers discuss water vapor and cloud changes separately, there can be little doubt that they are closely intertwined in the real climate system. After all, clouds can only come from water vapor. It might well be that getting the clouds in climate models correct requires getting the water vapor correct, and getting the water vapor correct requires getting the clouds correct.

If we examine the big picture, the average effect of all clouds is to cool the global climate system. This is because clouds, on average, reflect more solar energy back to space than they trap infrared energy. Despite the net cooling influence of clouds on the climate system, however, most climate modelers believe that clouds will change in ways that actually amplify global warming from greenhouse gas emissions, not reduce it.

Different kinds of clouds have different effects on the climate system. While all clouds shield the Earth from the sun, thin cirrus clouds high in the atmosphere end up trapping more infrared energy than they reflect solar energy. Thus, they have a net warming influence. An increase in high clouds would, by itself, warm the climate; a decrease in high clouds would, by itself, cool the climate.

In contrast to high clouds, low clouds almost always have a net cooling effect. This is because their solar reflection effect is stronger than their greenhouse warming effect. Many modelers now claim that low clouds represent the biggest uncertainty in global warming projections.

And even two clouds having identical altitudes, thicknesses, and water contents can still have very different effects on the climate. It turns out that the size of the tiny cloud droplets that make up clouds has a huge impact on how much sunlight a cloud reflects back to outer space. Many small droplets reflect much more sunlight than do fewer, large droplets. One real world example of this effect is the observation that the raining portion of a thunderstorm is typically brighter than the non-raining portion. The raindrops in the raining downdraft portion of the storm let much more sunlight through than the small droplets in the cloudy updraft part of the storm.

One of the hypothesized reasons for the global cooling trend between the 1940s and the 1970s is that manmade particulate pollution was causing clouds to form more easily, with more small droplets. A wide variety of tiny particles in the atmosphere, called cloud condensation nuclei, act as the point of initial growth of cloud droplets, and manmade particulate pollution is one source of these nuclei. It's a little ironic that some climate experts believe our cleaning up of atmospheric particulate pollution has made global warming worse. Maybe China's current economic boom and the resulting increase in air pollution will help us out in this regard.

Finally, there is a rather controversial theory that small fluctuations in the output of the sun can affect cloud formation. This effect, though, does not appear to be the result a change in total sunlight intensity, since satellite measurements of that intensity suggest very little if any change. Instead, there is a theory that the solar output of cosmic rays, as well as the shielding effect of the sun on the amount of galactic cosmic rays reaching the Earth, causes a change in the amount of cloud condensation nuclei. Laboratory experimental evidence supporting this controversial effect was first published in 2006. A minority of climate researchers believe that most of our current warmth can be blamed on long term variations in the sun. That sunspot activity is now at an historical high suggests that something might be going on with that big ball of fire in the sky.

So, how will clouds respond to increasing atmospheric carbon dioxide? This is widely agreed to be the largest source of uncertainty in climate model predictions of the future. Everyone agrees that clouds are a wild card in global warming predictions. A few scientists say that cloud changes associated with global warming might never be understood. Judy Collins was right, we "really don't know clouds at all."

Other Changes in the Climate System

There is a variety of other changes that can, at least theoretically, occur which would make global warming more or less severe. Sea ice, snow cover, and vegetation are a few of these. Sea ice and snow cover changes are widely believed to make global warming worse because, as warming progresses, the ice and snow melt which then exposes land or ocean surfaces that are darker than the ice and snow. These darker surfaces absorb more sunlight than they would have when they were covered by ice or snow.

Even less well understood are possible feedbacks from the biosphere. Changes in vegetation and oceanic microorganisms have the potential to either amplify or mitigate a warming tendency.

Finally, it seems like no one considers the possibility that some feedbacks might switch from positive to negative as warming progresses. It would not be that surprising for a complex fluid system like the atmosphere/ocean system to behave in such a way.

The Oceans

The world's oceans deserve special mention in any discussion of global warming because they cover over 70 percent of the Earth, and can hold over 1,000 times as much heat as the atmosphere. Like the atmosphere, the ocean is a fluid system that redistributes heat which originated as absorbed sunlight. But the huge thermal inertia of the ocean means that its temperature varies only slowly with any kind of change in solar or infrared radiation. This tends to moderate any temperature changes, minimizing them and spreading them over long periods of time. The time lag involved in the temperature response of the oceans is believed to be decades to centuries.

Like the atmosphere, the ocean is a chaotic nonlinear dynamical system that contains circulations of many shapes and sizes. Even without any help from mankind, unforced changes in ocean circulation are possible which could, at least theoretically, cause substantial changes in climate. For instance, a huge area of unusually cool water might surface, causing cooling of the overlying atmosphere for decades at a time. This might then change weather patterns, as they adjust to a different geographic distribution of heating. Since we do not understand why these changes in ocean circulation occur, the ocean represents still another large source of uncertainty in our efforts to discern any human influence on climate.

The ability of the oceans to cool the atmosphere is dramatically demonstrated off the west coasts of North and South America. In these areas, the general circulation of the ocean is characterized by the continual upwelling of cool waters from below, as well as transport from the polar regions to lower latitudes. These cool surface waters then flow slowly westward in the tropics and subtropics, gradually warming from absorbed sunlight. Eventually, some of this heat is circulated back poleward in the western Pacific, helping to warm the polar regions. There the water cools, and the cycle starts all over again. Like the atmosphere, the most fundamental function of the ocean circulation is to transport heat from where there is more to where there is less.

One curious, but seldom mentioned observation is that most of the water in the ocean, even in the tropics, hovers around the frigid temperature of 40° Fahrenheit. It is only a relatively shallow layer of water in the upper part of the ocean that warms up from absorbed sunlight. Cold, deep-ocean water is believed to be continually replenished in the Arctic Ocean, where surface waters become so dense from saltiness that they sink. Might this huge reservoir of cold water represent a long term buffer against warming? No one knows.

A More Optimistic View of Global Warming

All climate models exhibit net positive feedbacks, that is, they magnify the small amount of direct warming from mankind's production of greenhouse gases. Most skeptics I know, in contrast, believe that the climate system will act to reduce, not amplify, that warming. The modelers then retort that their predictions of warming could just as easily be an underestimate as an overestimate.

This view, however, makes it sound like the processes at work in the climate system are just some random combination of positive and negative numbers. Instead, I believe it makes more sense to assume that the Second Law of Thermodynamics is the ultimate guiding principle in climate sensitivity, and that the climate system changes in ways that act to rid the system of excess heat.

Some say that the fact that most, if not all, climate models predict substantial global warming should be reason to have confidence in them. But there are other reasons why different climate modeling groups would get similar, but wrong, results. They all draw upon the same (but incomplete) body of published research that their models are built from.

Furthermore, there is unspoken peer pressure between the different modeling groups to "fit in": the modeling group with the most or least warming feels some pressure to conform to the rest of the group. As one of our leading climate model experts, Bob Cess, admitted to *Science* magazine in 1997, "the [models] may be agreeing now simply because they're all tending to do the same thing wrong. It's not clear to me that we have clouds right by any stretch of the imagination." While there has been ten years of progress since then, the range of warming forecasts produced by even the latest, state-of-the-art models has narrowed only slightly.

I would like to suggest an alternative, more optimistic view of global warming. In this view, warming will be relatively benign because the climate system tends to stabilize itself against the warming influence of increasing greenhouse concentrations. There are some good reasons to believe that this alternative view has merit.

First, we note that the amount of absorbed sunlight, emitted infrared heat, and strength of the greenhouse effect are not self-existent static quantities. They are largely under the control of weather processes. Weather systems determine how much of the available sunlight the climate system will use before clouds, in effect, turn off the solar energy spigot at about 80 percent of what could be used.

Similarly, weather systems determine how much greenhouse effect (mostly water vapor and clouds) they will produce and maintain.

There are many other combinations of clouds, water vapor, absorbed sunlight, and emitted infrared radiation that would also satisfy global radiative balance and produce a relatively constant globally averaged temperature. Why does the climate system choose only the one combination that we see, year after year, producing average temperatures that are stable to within a few tenths of a degree? Climate modelers, in effect, fiddle with all of the different "control knobs" in their models to try get the correct average amounts of clouds, water vapor, and temperature. The climate system knows exactly why it has chosen to maintain those averages. It is not at all obvious that the models that get close to the right average climate do so for the right reasons.

In my view, the mere existence of a preferred climate state over other possible states is evidence of climate stability. Indeed, historically there has been a tendency for climate models to wander away from an average state. This is called model "drift," and represents empirical evidence that the climate models are, if anything, too sensitive to changes imposed on them.

The "control knob" settings that have been chosen for one amount of atmospheric carbon dioxide might not be the same ones that are needed for increased carbon dioxide levels. We simply don't know which, if any, should be changed.

THE EARTH'S THERMOSTAT

I believe that the available evidence suggests that the Earth has a thermostatic control system—but unlike the one pictured at the beginning of this chapter, it isn't mounted on a palm tree on some deserted island. The real control system is precipitation.

When we reflect upon the internal processes that control the Earth's natural greenhouse effect, cloud amounts, and temperature, we find that almost all of them can be traced to the behavior of precipitation systems. These systems determine how much water vapor, our main greenhouse gas, is removed from the atmosphere, and thus how much will be left to cover the Earth.

These systems influence cloud amounts and types through their control of the vertical temperature structure of the atmosphere. Remember the illustration in the last chapter that showed how dramatically the vertical temperature structure is changed by weather systems? This vertical temperature structure then indirectly helps control the formation of most clouds—even clouds far away from any precipitation activity.

Clouds, in turn, determine how much sunlight will be allowed to reach the surface, and they also constitute the second strongest portion of the Earth's greenhouse effect. See how intertwined everything is in the climate system?

Even the low stratus and stratocumulus clouds that form over the subtropical oceans, thousands of miles from any precipitation activity, are there because of precipitation. The clouds form from moisture being trapped beneath a temperature inversion (warm air layer). That inversion is, in turn, caused by air being forced to sink in response to rising air in precipitation systems.

A few climate researchers I have talked to don't understand how precipitation, which covers only a few percent of the Earth at any given time, could have such a controlling influence on our climate. They do not appreciate the fact that the air we are breathing right now was inside precipitation systems some number of days or weeks ago, being recycled for the umpteenth time, so that it could do its job of maintaining the climate system at its preferred state.

Think of the thermostat and air conditioning system in your house. It occupies only a small portion of the house, yet we know that we can't hope to understand how the temperature in the house is controlled unless we understand the thermostatic control mechanism. The same is true of the atmosphere. Precipitation systems might be small, but they exert tremendous influence. While humanity slowly pumps relatively tiny amounts of carbon dioxide into the atmosphere, precipitation systems are spewing vast quantities of water vapor out of their upper levels, adjusting the amounts up and down to help maintain the climate system at what appears to be a preferred average temperature.

It seems that much of nature operates in this way, with built-in checks and balances. When the system veers too far from normal, complex processes interact in ways that push the system back in the opposite direction. Scientists and environmentalists who believe in an unstable climate system, in effect, do not believe that these restoring forces exist. They believe that if the system is pushed away from its average state (say, by increasing carbon dioxide concentrations), that it continues to travel in that direction, amplifying the initial tendency, possibly even taking the climate system past some imagined tipping point from which there will be no return.

Since previous theoretical work has indicated that an increase in precipitation efficiency results in a cooler climate with (somewhat counterintuitively) *less* precipitation, we know that the climate system has at least this one thermostatic control mechanism at its disposal—if it chooses to use it. If such a mechanism is indeed in operation, the net effect of manmade greenhouse gases on the climate system could conceivably be so small that we wouldn't even be able to measure it. A small increase in precipitation efficiency in response to the warming tendency from extra carbon dioxide could potentially result in no measurable change in either temperature or precipitation. Most of the changes will have occurred inside precipitation systems, where the mysteries of climate sensitivity are kept hidden from our weather sensors.

The foregoing examples and arguments represent one possible climate stabilizing mechanism which I present as a hypothesis. It is not entirely original, as it builds upon the published work of others. While far from being proved, it is still consistent with much of our conceptual understanding of how the atmosphere operates. I would expect that there are other possible climate stabilization mechanisms as well which I have not considered here.

It could be that science has only scratched the surface when it comes to knowing how the climate system regulates itself.

So, What Do We Really Know About Global Warming?

Given all of the uncertainties, what can we say that we really know with reasonable certainty about manmade global warming? First, we know that mankind is producing carbon dioxide as a result of our use of a wide variety of fuels, from coal and petroleum to natural gas and wood. No scientist that I know of disputes this.

A second observation we are certain of is that the carbon dioxide content of the global atmosphere has been slowly increasing. We are now about 40 percent of the way to a doubling of the pre-industrial concentrations of atmospheric carbon dioxide. While this might sound dramatic, today's carbon dioxide concentration is still very tiny, amounting to only 38 molecules of CO_2 for each 100,000 molecules of air. To those 38, mankind is adding about 1 molecule of CO_2 every five years or so.

There are a couple of important points that need to be made about this CO_2 increase. It is unlikely that the increase is due to natural processes, since mankind's burning of fuels produces more than enough carbon dioxide to explain the observed increase. Only about 50 percent of what we produce ends up staying in the atmosphere. The rest is "missing"—presumably absorbed by the ocean and the biosphere, fertilizing plants and thus increasing vegetation growth rates around the world. Note that, since this aspect of human "pollution" is actually good for vegetation, you won't hear many environmentalists mention it. We wouldn't want anyone to get the impression that some aspects of mankind's carbon dioxide emissions might be beneficial.

You might sometimes hear that the Earth itself emits much more carbon dioxide than humans do. While this is true, it is misleading. Huge amounts of CO_2 are indeed continuously being both emitted and absorbed by the oceans and land, but presumably the amounts emitted by the "sources" have been in balance with the amounts absorbed by the "sinks." That is, the system is assumed to have been in balance. The prevailing opinion now is that today's gradual increase in the atmospheric concentration of CO_2 is evidence that humans constitute the new, additional source of carbon dioxide.

Thirdly, we know that carbon dioxide is a greenhouse gas, which means that it traps infrared radiation and so tries to warm the lower troposphere to a higher temperature than if the gas was not there. But as was demonstrated in the last chapter, the cooling effects of weather have a stronger influence on surface temperatures than the warming influence of greenhouse gases.

Finally, we are pretty sure that the globally averaged surface temperature of the Earth is at least 1° Fahrenheit warmer now than it was about a century ago. As much as 40 percent of the increase occurred before 1940, which cannot be entirely blamed on greenhouse gas emissions simply because mankind had not used very much fossil fuel up to that point. The rest of the increase has occurred since the 1970s.

So, carbon dioxide concentrations have risen, and globally averaged temperatures have risen. But to what extent has the first caused the second? Coincidences do, after all, exist—that's why we have a word for them.

It is very difficult to confidently attribute the current warming to a specific cause or causes. And that is what makes the claim that global warming is due to humans more of a belief system than a scientific observation. Attributing most or all of the current warmth we are experiencing to mankind is a statement of faith, because it assumes something we don't know: how much natural climate variability there has been during the same period of time.

One of the reasons why so many scientists believe current warmth is due to human emissions of carbon dioxide is simply due to the fact that the human emissions are known, while natural climate variability is, for the most part, *unknown*.

Without knowing how much of the current warming is due to natural variability, there is no way to know to what extent mankind is involved in climate change. For instance, there has been a substantial decrease in Arctic sea ice during the summer melt season in recent decades, and Arctic temperatures have risen faster than anywhere else. Scientists and environmentalists point to this fact as a sure sign of manmade global warming. But the available thermometer measurements in the Arctic region suggest that it was just as warm back in the 1930s, which couldn't have been caused by the activities of mankind. What were Arctic sea ice conditions like back then? We don't know because we've only been able to monitor that remote region with satellites since 1979.

Even the 1° Fahrenheit warming in the last century is somewhat uncertain. Many long-term thermometer measurements are known to have spurious warming due to the "urban heat island" influence. This is the tendency for the micro-climate near manmade structures to slowly warm over time as natural vegetation is gradually replaced with buildings, parking lots, sidewalks, etc. The effect is strongest in major metropolitan areas, with average temperature increases of several degrees Fahrenheit.

Those who analyze the thermometer data claim to have corrected for these urban heat island influences. This is done by comparing urban sites to more rural thermometer sites. What cannot be adjusted for, however, is any warming of the rural sites which have experienced an increase in manmade structures, too. After all, people like to build things. The addition of an outbuilding near the thermometer site or a change in land use (for instance irrigation) can cause a spurious warming effect that might be blamed on carbon dioxide. I believe that this residual urban warming effect, which will probably never be accurately known, accounts for a small warming bias that still remains in the thermometer record of global temperature trends.

In summary, the things we are pretty sure of are: 1) carbon dioxide concentrations in the atmosphere are slowly increasing, very probably due to humans, 2) carbon dioxide is a greenhouse gas, and 3) globally averaged temperatures have risen by about 1° Fahrenheit since about 1900. All of these facts taken together are certainly consistent with the hypothesis that mankind is causing global warming. But we should keep in mind that the first and only theory of some newly observed phenomenon seldom survives forever.

The Future of Global Warming Research

In spite of their shortcomings, climate models remain our only way quantitatively to estimate global warming and its effects on the climate system. The models are undergoing continual changes, with improved understanding of a wide variety of physical processes that combine to form what we call weather and climate.

But future success is not guaranteed. As discussed previously, the unpredictable behavior of nonlinear systems like the atmosphere and ocean can thwart our attempts to model these complex systems. We might be able to get realistic model behavior for the subsystems, but then the realism can disappear when the subsystems are put together to model the whole system. Better building blocks do not guarantee better global warming predictions. Nevertheless, the importance of the global warming issue demands that we try.

Maybe it is time to entertain alternative paradigms to climate modeling. The traditional approach has been to assemble the model of the whole system from its component subsystems, which in turn have their own sub-subsystems. This bottom-up modeling approach has always been favored in the physical sciences. But there is an alternative approach, one which has become more popular in nuclear physics. That is to step back and examine the overall "emergent properties" of the climate system. In this approach, the major macro-scale features of the climate system are explained in as simple terms as possible, with as few variables as possible. Additional features are added to the model only when they are necessary to explain the overall behavior of the climate system. By looking at the climate problem in this way, it might be possible to arrive at a better idea of how sensitive the climate system is to increasing levels of carbon dioxide.

Such an approach is likely to require a more complete exploitation of satellite data, which provides our only way to observe the behavior of the whole climate system on a global basis. Of course, it is mere coincidence that analysis of satellite data is my specialty. Still, I would be more than happy to accept a research grant to look into this problem, if you wish.

We've put the exhaust pipe on the inside!

Chapter 5: The Scientists' Faith, the Environmentalists' Religion

What scientists claim to know about manmade global warming is based as much upon faith as it is upon knowledge. The climate modelers have faith that they understand the mechanisms that control the climate system. They have faith that their mathematical representation of those processes in the models will cause the models to behave realistically. Might the models be correct? Sure. But as the last chapter demonstrates, the evidence is not as compelling as you might have been led to believe.

And you don't have to just take my word for that. Results of a survey of 530 climate scientists released by The Heartland Institute in May of 2007 revealed that only about one-half of the scientists agreed that "climate change is mostly the result of anthropogenic (man-made) causes." Only one-third agreed that "climate models can accurately predict climate conditions in the future." That doesn't sound like the science is settled to me.

The fact that emotions run so high on the subject of global warming is a sure sign that more than just science is involved in the debate. Science doesn't care what the answers are to our questions about climate. Borrowing from a previous example of mine, would there be this level of interest if the debate was over two different scientific opinions on the mating behavior of the tsetse fly? I don't think so. The roots of the conflict over global warming go much deeper than a simple disagreement over what's happening in the climate system.

One piece of evidence that manmade global warming is still just a theory is that scientists will talk in probabilistic language about it. For instance, a scientist might state that he thinks there is a 90 percent chance that most of the current global warmth is due to mankind's burning of fossil fuels. But as a statistician will tell you, probabilities are either based upon a lot of data collected of past events, or they are based upon a known number of possibilities, like the probability of rolling snake eyes with a pair of dice (1 in 36). Global warming is an example of neither. For manmade global warming, there is only one event, and it is, or is not, occurring right now. So probabilistic language applied to global warming is misleading and inappropriate. Its use is simply a pseudo-scientific way of conveying the level of faith a scientist has in his beliefs.

There is no question that scientists bring preconceived notions and biases along when they perform global warming research. In the absence of proof, scientists fall back on their intuition, or on past experience in their specific field of expertise.

Climate Modelers Versus Meteorologists

It is natural for scientists to put undue trust in their own research. After all, their livelihoods and reputations are at stake. And in the case of climate modeling, a large group of individuals from different specialties and having different talents have invested many years in building and improving each climate model. It is understandable, then, that as the model gradually becomes better at imitating the average behavior of the climate system, the modelers tend to believe the global warming that the model produces.

Over the years, I have noticed a distinct difference in the way climate modelers and meteorologists perceive the climate system. Climate modelers are usually physicists who are typically better at computer modeling than meteorologists. Physicists are more accustomed to reducing the behavior of a physical system to a minimum number of mathematical equations in order to study it.

But physicists tend to have a simpler view of how the weather works than do meteorologists. They usually have little or no formal education in meteorology. In contrast, we meteorologists appreciate the inherent—almost biological—complexity of the climate system. Based upon our experiences with weather forecasting and watching the weather, we view the climate system as being self-regulating.

As a result of the difference in backgrounds between climate modelers and meteorologists, I find much more skepticism about global warming among meteorologists than among the physicists/ modelers. I believe this is just one more reason why modelers are often unduly confident in their model predictions.

Real Scientists Don't Say "I Don't Know"

The fact that scientists are human with their own biases (Chapter 2), combined with the uncertain nature of global warming theory (Chapter 4), leads to a tendency for climate researchers to portray their research findings to others in a more definitive way than they should. In other words, they don't know as much as they claim to know. This is especially true when being interviewed by a reporter. After all, we're the ones that are supposed to know. We are the experts, the rocket scientists.

Maybe we think it looks like we aren't doing our job if we sound uncertain. Or maybe we worry that we will sound wishy-washy, or uninformed. Or maybe we're afraid of losing our government funding if our research isn't viewed as successful. All of these reflect basic human tendencies that inevitably affect how scientific research results are reported.

And what about scientific uncertainties arising from things those scientists are not even aware of? While we can address uncertainties in science that we know about, we must take on faith that there are not important processes at work which are, as yet, totally unknown to us. The climate system probably still has several such surprises in store for us.

Some climate scientists act like they are doing something worthwhile for humanity by expressing alarm about global warming. Just like most environmentalists, they seem to think that risks should be reduced through more government regulation. In rare moments of openness, a few of the more outspoken believers in dangerous global warming have admitted to this bias. Several have said, in effect, "Even if global warming isn't going to be a problem, reducing fossil fuel use is the right thing to do anyway."

Stephen Schneider has often been either misquoted, or at least quoted out of context, regarding his views on the public service role of the climate scientist. Some have claimed that he suggested that climate scientists need to be dishonest in order to be effective in facilitating change. In the interests of accuracy, I reproduce his entire quote from an October 1989 *Discover* magazine article by Jonathan Schell. While not advocating dishonesty, Dr. Schneider's views are nevertheless very revealing.

> On the one hand, as scientists we are ethically bound to the scientific method, in effect promising to tell the truth, the whole truth, and nothing but—which means that we must include all the doubts, the caveats, the ifs, ands, and buts. On the other hand, we are not just scientists but human beings as well. And like most people we'd like to see the world a better place, which in this context translates into our working to reduce the risk of potentially disastrous climatic change. To do that we need to get some broad-based support, to capture the public's imagination. That, of course, entails getting loads of media coverage. So we have to offer up scary scenarios, make simplified, dramatic statements, and make little mention of any doubts we might have. This "double ethical bind" we frequently find ourselves in cannot

> be solved by any formula. Each of us has to decide what the right balance is between being effective and being honest. I hope that means being both.

Professor Schneider, a biologist, does not appear to be advocating outright dishonesty. Instead, for the sake of "making the world a better place," he does suggest that scientists should exaggerate the level of confidence we have in the possibility of disastrous climatic change.

As I will address in the next chapter, this unnecessarily escalates the perceived risks of climate change, which then can lead to biased policy actions when it comes time to weighing risks against benefits.

While such actions might be born of good intentions, they can result in damage to human interests when the unintended negative consequences of misguided policy actions start appearing.

The public has become appropriately skeptical of scientists' predictions of environmental doom. After all, scientists' batting average in this regard has been close to zero. For instance, early estimates of global warming back in the late 1980s were, in retrospect, too large. Paul Ehrlich's *Population Bomb* bombed on its forecast of widespread famine. And while there is some small risk associated with the use of DDT, knee-jerk bans on the insecticide have killed literally millions of Africans.

Some will claim that the citizen has no right to distrust the consensus view of scientists. After all, the citizen is no expert. Well, the average citizen doesn't have to be an expert in some field of science to know that the outlandish predictions of even world renowned experts are likely to be wrong. While I might not be an expert on the inner workings of the human mind, if a brain researcher tells me that he can take a few measurements of my brain activity and then tell me what I will be thinking exactly twenty-four hours from now, I'll still wager that he will be wrong.

There is a truism in science that the more we learn from the scientific investigation of a subject, the less it seems we understand. At first, a hypothetical explanation for some newly studied physical process might seem remarkably clear. Our understanding is simple, uncomplicated by details. Then, the more data we collect, the more muddled things become. We find exceptions to what we thought were rules. We find that the process we are studying also depends upon other factors we didn't think of.

It takes a higher level of understanding to appreciate the fact that for every change in a natural system, there are a variety of responses, each of which needs to be understood to explain the behavior of the total system. I suppose it is only human nature to approach a problem with the expectation that it won't be that difficult to solve. It is similar to when I plan a home improvement project. I *always* underestimate the cost and time required to complete it.

That the Earth's climate system is possibly the most complex physical system we know of should, by itself, humble the climate scientist. What is written about nonlinear dynamical systems usually deals more with how marvelous and complicated they are, rather than with how to predict them. The deeper we probe, and the more we learn, the less we understand, and so the more amazed we are at how nature works.

Scientists have worked for many years to model the behavior of the climate system with computer codes. Many millions of dollars have been spent trying to build models that are sufficiently accurate to give us a good idea of how much warming we might experience in the coming years and decades. And the modelers would like to be able to say that these efforts have been a success. Certainly great progress has been made, but no matter how much we spend on the modeling effort, or how hard we try to predict future climate, there is no guarantee that we will be successful.

While computer modeling is probably our only hope for forecasting climate change, the push for success typically leads to an overstatement of how well the computer models work. Ultimately, how much you choose to believe climate model predictions depends upon your faith that the models contain the most important processes to predict future climate.

There is an interesting dynamic that occurs when climate

modelers get together to compare their results. Out of one or two dozen modeling groups around the world, no one wants to be an "outlier," for example the group that gets the least warming or the most warming from their model. You could call it scientific peer pressure to conform. Rather than objectively analyzing the possibility that the best model might indeed be one of these outliers, there is instead a subconscious and unspoken pressure to conform to the average of all of the models.

In statistical terms, what this means is that the model errors are assumed to be *random*, in which case the average of all the models would be much closer to the truth than any particular model. But the most important errors are more likely to be *systematic*. This means that all of the models have a bias in one direction because they all are missing one or more important process. Scientists don't like to talk about this possibility very much because we cannot study things we do not know about.

This is just another example of the scientist's tendency to forget that his results are dependent upon his assumptions being correct. Some of these assumptions were knowingly made in order to simplify the problem. For instance, climate models are purposely simplified so that they can run to completion on today's computers and give the researchers results before they reach retirement age. Believe it or not, computers are still nowhere near fast enough to run a climate model with all of the processes we know about, explicitly represented, and in high definition. Even more assumptions are unknowingly made due to processes that we are not yet even aware of.

You would think that scientists routinely question their assumptions, or at least make it clear that their explanations for how things work are based upon their assumptions being correct. Unfortunately, this is seldom the case. It is no accident that the biggest advances in science are usually the result of someone going back and questioning what had been assumed for many years to be true. And it would be hard to imagine a scientific endeavor with more assumptions, explicit and implicit, than climate modeling.

There is an old saying, "to someone with only a hammer, everything looks like a nail." Climate scientists tend to view everything in the context of global warming theory. We know humans produce greenhouse gases and pollution aerosols. We know these must be having some effect on the climate system. We come up with quantitative estimates their effect and insert them into the models.

In contrast, we do not understand what causes natural fluctuations in climate, so we tend to ignore what we do not understand. Global warming theory is our hammer, and we have a tendency to explain all changes (nails) that we see in the climate system in the context of that theory.

Another analogy is looking for something you lost on a dark street at night. If the only place you can actually see is under a street light, then that is where you will look. Global warming theory is our street light, illuminating only a small part of the climate change problem. It is the only place where we look for explanations because it is the only place where we can see clearly. We understand that humans produce carbon dioxide, and that carbon dioxide is a greenhouse gas. Since this potential source of climate change is the only one we really understand, we have a tendency to see evidence of it everywhere.

I've had discussions with a couple of climate modelers about what I believe to be remaining uncertainties in the way the climate system works. They both finally declared the same thing: the climate has warmed; what else could the explanation be other than mankind's production of greenhouse gases? To me, this attitude is evidence of a strong faith that there are not natural climate variability processes at work that might also explain the warming.

I am not claiming that we should wait until we fully understand the climate system before we make policy decisions, because that level of understanding will never be reached. In fact, in Chapter 9 (Less Dumb Global Warming Solutions) I will describe how we are already doing much of what is necessary to reduce the risk of future warming.

Instead, I am claiming that our confidence in current climate models' ability to predict the future is misplaced. This is partly due to the inherent complexity of the climate system. But it is also due to the tendency for scientists to project overconfidence to the outside world. Apparently, since scientists are supposed to have all of the answers, real scientists don't say "I don't know."

Climate Change Denial

The overconfidence of some scientists, along with the underlying motives of environmentalists, politicians, and the media, is leading to some impatience with the views of global warming skeptics. As a result of the apparent nobility of their mission to spread the truth of imminent global warming catastrophe, some environmentalists and reporters have started to demonize those who would dare to disagree with them. At this writing, a Google search for the phrase "climate change denial" now returns over 70,000 web pages.

Calling someone a global warming denier implies that global warming skeptics do not believe in global warming. But this charge is completely false. I know of no skeptics who deny that global warming has happened. In this way, those whose agendas are so important to them that they cannot simply let the facts speak for themselves are now resorting to intimidation, under the guise of "good science."

Extreme statements like "all reputable scientists believe in global warming" are, at best, misleading. At worst, they are propaganda. The purpose of such statements is to cast *ad hominem* insults in an effort to discredit others with opposing viewpoints. If you don't agree with the majority of scientists, you are a crackpot, or a denier of the Holocaust or the dangers of tobacco. This is a favorite technique in the propagandists' bag of tricks.

How could anyone dare to question the authority of the Consensus of Scientists? This current trend toward equating climate change denial to some of the most indefensible beliefs of our time would be humorous if it wasn't so dangerous. The media have gleefully participated in this blanket condemnation of those of us that have the audacity to believe that the climate system is not on the brink of disaster.

For instance, when asked why opposing viewpoints were not included in a *60 Minutes* special on global warming, the interviewer responded, "If I do an interview with Elie Wiesel, am I required as a journalist to find a Holocaust denier?" One Australian commentator has actually suggested that it might be time to outlaw climate change denial.

Al Gore has also begun to play this game, stating, ""Fifteen per cent of the population believe the moon landing was actually staged in a movie lot in Arizona and somewhat fewer still believe the Earth is flat. I think they all get together with the global warming deniers on a Saturday night and party." That would be pretty funny if the consequences were not so great.

Grist, an online environmental news and commentary magazine, posted the following comment by one of its writers concerning those who would stand in the way of global warming alarmism:

> When we've finally gotten serious about global warming, when the impacts are really hitting us and we're in a full worldwide scramble to minimize the damage, we should have war crimes trials for these bastards—some sort of climate Nuremberg.

This is an amazing point of view, and it illustrates the level of emotional attachment many people have to fears of environmental disaster.

If we are going to play the blame game, then I would like my turn now. I believe that the environmentalists who have stood in the way of allowing the use of DDT in Africa are the real criminals. Rather than some theoretical future threat like global warming, the DDT ban is now known to have needlessly cost millions of lives in Africa. Where is the outrage over this very real tragedy? Is the silence because we don't really care about what happens to dark-skinned people in poor countries? The western world's adherence to our secular religion, environmentalism, is apparently more important to us than the unnecessary deaths of millions of black Africans.

Usually noncommittal in areas of scientific debate, scientific organizations have started taking sides on the global warming issue. The Royal Society of London has written a letter to ExxonMobil demanding that the oil giant stop funding global warming skeptics. Despite the Royal Society being a scientific body, the letter openly mentions the negative effect that the skeptics have on the adoption of the Kyoto Protocol to reduce greenhouse gas emissions. As I will discuss in the next chapter, this misguided foray into economics and policy advocacy only serves to illustrate the Society's political biases.

These tactics should be an affront to those who claim to defend free speech. Or is free speech only defended when it supports a particular political or environmental agenda? If the claims of global warming skeptics are so ridiculously wrong, why not simply let us speak and thereby make fools of ourselves? Why not put our claims in the spotlight, and under the microscope, and show everyone the obvious stupidity of our positions?

Global Warming as Religion

One of the definitions for religion you will find in Webster's dictionary is "a cause, principle, or system of beliefs held to with ardor and faith." While global warming is a legitimate area of scientific study, those who believe in the catastrophic view of manmade global warming might best be described as religious disciples. For them, human interference in the climate system is evil. Without mankind the Earth would be undefiled. Our use of natural resources is a transgression against our Earth Mother.

In contentious issues, one finds that people generally believe what they want to believe, rather than what the evidence leads them to believe. This can make rational discussion of global warming issues very difficult. Holding certain beliefs about the natural world and the climate system is fine. The trouble I have is with those who try to pass such beliefs off as being "science." Global warming being "bad" might be a philosophical or religious belief, but it is not scientific. As I mentioned before, science doesn't care whether the Earth is warming, cooling, or staying the same. Only people care.

How else could we categorize these beliefs, other than religious? Does the Earth have some divine right to remain untouched by mankind? Why are all other parts of nature allowed to influence the climate system, but not humans? Mount Pinatubo can spew millions of tons of sulfur into the stratosphere, and that is considered part of nature. If mankind does it, we are destroying the Earth. What's wrong with this picture?

Despite constitutional prohibition against the favoring any specific religion, we are now teaching our schoolchildren to repent of their sins against nature. In contrast, we never hear of students being taught of the benefits of global warming. For instance, we now know that the extra carbon dioxide and global warmth, no matter what their cause, are resulting in a gradual greening of the Earth. There is some evidence that there has been a slight poleward shift in the habitats of some warm weather species, from the tropics where there is greater diversity of life, towards higher latitudes where many of these forms of life could not otherwise survive. Global warming has made winters less severe, and cold weather is known to cause more deaths than hot weather. So why is global warming necessarily a bad thing?

I'm sure if mankind was accused of global cooling and causing a destruction of carbon dioxide (which is food for vegetation) there would be howls of protest that we are strangling the biosphere. So, why are we not hearing the environmentalists applauding humanity for helping to create more life?

Everything in nature affects the climate, and the climate affects everything in nature. If the greatest diversity of life is found in the tropics, might it not be a good thing if the tropics expanded slightly to cover a little bigger area? Why is it that a forest affecting climate is good, but humans affecting climate is bad? Why is it that the longer growing season is cited as a negative, instead of a positive, impact of global warming on humanity?

Apparently, the state of the climate system in 1967 (or pick any other year) was "optimum." Any deviation from this is bad, unless that change was caused by some other part of nature and not mankind. The Earth apparently has a "right" to be free of human influence.

But "rights" are a uniquely human construct. I hate to sound harsh, but forests and other forms of life on Earth have no rights, except for the ones that humans might want to confer upon them. I can just imagine one species of fish deciding to stop eating another species of fish because it wants to respect their rights. We all love polar bears, but why do they continue to infringe on the rights of seals?

Please don't think I'm anti-environment. I believe that it is a good thing to preserve some old growth forests, to minimize water pollution, to conserve energy. I enjoy the fox that lives in my backyard and the deer that come to visit. I don't hunt wild animals because I no longer enjoy killing something just for sport. But these elements of nature have value only because humans value them, not because the environment has some basic right to remain undisturbed.

The religious reverence some have for the environment is probably best categorized as Paganism. While there are many variations in Pagan beliefs, they typically involve the Earth, life, and all the cosmos being part of one spiritual being. For instance, Al Gore's first book, *Earth in the Balance: Ecology and the Human Spirit*, addressed the spiritual connection that Mr. Gore has with the environment. The theme of that book is anti-technology, and mankind is religiously viewed as a destroyer of the environment rather than as a species that happens to depend upon a wide variety of natural resources for it to thrive. While Mr. Gore is a Baptist, some of his writings and speeches to environmental groups sound more Pagan than Christian.

I don't want to sound like I'm demeaning Pagans, because I'm not. As a religious belief system, it is peaceful and optimistic; it recognizes the marvelous complexity and interconnectedness we see in nature. I'm only trying to point out that such beliefs regarding humanity's relationship to nature are inherently religious. I'll leave it to you to decide whether such beliefs should be taught in public schools.

As is the case with more traditional religions, there are those who are nominal believers, and those who are ardent believers. We usually end up hearing from the true believers at some point through the media. That popular periodical of newsworthy scientific discoveries, *Science*, ran an article in 1967 in which Lynn White, Jr., a U.C.-Berkeley professor, stated

> More science and more technology are not going to get us out of the present ecological crisis until we find a new religion, or rethink our old one.

In 1982, the founder of Greenpeace, Paul Watson wrote of his religious environmental awakening:

> I got the impression that instead of going out to shoot birds, I should go out and shoot the kids who shoot birds.

The belief that global warming is a serious threat to mankind and the environment has been described as having a striking similarity to the biblical paradigm of sin, guilt, and the need for redemption. The author Michael Crichton has done a brilliant job of articulating the modern secularist's subconscious need for religion in his life, as evidenced by the secularist's reverence for the environment. In a 2003 speech, Crichton summarized these parallels between modern environmentalism and the Judeo-Christian belief system:

> There's an initial Eden, a paradise, a state of grace and unity with nature, there's a fall from grace into a state of pollu-

> tion as a result of eating from the tree of knowledge, and as a result of our actions there is a judgment day coming for us all. We are all energy sinners, doomed to die, unless we seek salvation, which is now called sustainability. Sustainability is salvation in the church of the environment. Just as organic food is its communion, that pesticide-free wafer that the right people with the right beliefs, imbibe.

Crichton makes it clear that he thinks we should be good stewards of the environment, but that emotion often gets in the way of facts when it comes to making decisions about what being "good stewards" means from a practical standpoint. He also points to the example of the international bans on DDT, actions that were based more on emotion than sound science. DDT, by itself, would greatly alleviate the scourge of malaria in poor African countries, with almost no risk to humans or wildlife. Instead, millions of people, especially children, continue to die from this largely preventable disease.

Might this be part of the environmentalists' religious rites? They offer sacrifices of children—but only children in some far-off land who have a skin color different from most of us. A famous climate researcher was once giving a talk at a conference, and someone in the audience brought up the potential for many deaths in India caused by severe weather. To everyone's astonishment, the scientist said something to the effect of "you make it sound like millions of Indians dying would be a bad thing."

The Gaia Hypothesis

The Gaia Hypothesis was introduced by James Lovelock and Lynn Margulis (formerly the wife of the famed astronomer and writer Carl Sagan). It is the modern pseudo-scientific resurrection of ancient Pagan beliefs about the universe being one with some spiritual entity. Specifically, the Gaia Hypothesis views the biosphere as a single living being, Gaia, whose name is taken from

Greek mythology. I'm not quite sure how Gaia engages in reproduction, but I'm pretty sure I don't want to be around when it happens.

In today's scientific age, the Gaia movement and its variants would seem to be the logical alternative to more traditional religions for those who have spiritual needs, but who don't want the inconvenience of moral demands on their character. I have found that a number of scientists have this as their religious belief system. When NASA began its Earth system science research program many years ago, a periodical of research progress was started that was given the name "Gaia." The name didn't last long, though. I suspect someone pointed out that the religious connotation might be a little over the top.

Global warming is considered to be the ultimate sin against the Earth. Mankind is giving Gaia a fever, and she is getting pretty irritated about it. Like the human body that fights off an infection with a rising temperature, global warming is Gaia's way of ridding herself of this infection called "humanity." It is easy to see why this kind of belief has such wide appeal.

ENVIRONMENTALISM IN THE CHURCH

Many Bible-believers have now bought into the catastrophic view of mankind's influence on nature. Environmentalists have done an admirable job of enlisting Christian and Jewish organizations to help reach their goals. After all, the Bible has several passages which suggest that we should be good stewards of the creation. The first instance was God's command to Adam to tend the Garden of Eden. From this point of view, it seems appropriate that people of faith would be involved in efforts to care for the environment.

But the first book of the Bible, Genesis, also tells Adam to "fill the Earth, and subdue it." Some have blamed the wanton exploitation of the environment by mankind on this passage. These seemingly contradictory instructions to Adam have led to a certain level of tension within churches and synagogues, since exactly what is meant by being "good stewards" is open to a wide range of interpretations. And it might well be that the biblical intent is for some level of ambiguity on this point, allowing people the freedom to decide what stewardship means for themselves.

As is often the case, different denominations decide to emphasize some scriptural issues more than others. Perhaps predictably, some leaders of the Christian environmental movement have decided to err on the side of the environment, rather than mankind. The Cathedral of St. John the Divine, in New York City, has had a history of performing services that verge on Earth-worship. At times, their Pagan teachings have centered on the oneness of the universe, the universe as God, and Gaia as our Earth Mother.

The Bible, in contrast, makes it clear that the Creator is separate from His creation, and warns against those who would "worship the creation, rather than the Creator." For those who have a spiritual need to worship something, ultimately these are the only two choices; there is no third. You either worship the Creator, or the creation. The atheist believes that matter is the ultimate reality, and in effect worships matter's physical laws. The Bible-believer believes that the spirit of God is the ultimate reality, and so worships the Giver of both physical and spiritual laws. The agnostic doesn't really care enough to choose one over the other, and maintains that you can't really know for sure anyway.

The National Religious Partnership for the Environment (NREP) has grown out of the New York movement to be the dominant environmental organization within the church. NREP mailings have been received by over 60,000 churches in the United States, and many denominations have signed on as partners. I suspect that many, if not most, of these churches are unaware of some of NREP's environmental extremist leanings. Churches have organized workshops, recycling activities, classes on our responsibility to the environment, and a host of other activities for their members.

Some church leaders have considered the issue so critical that they have taken political sides in elections, and voiced support for this or that policy change to address the global warming threat.

But, as I will explore in the next two chapters, it can be counterproductive when a church adopts the policy preferences of environmentalists—it can even be contrary to the church's stated mission. The church has not been told the truth about the negative, unintended consequences that will result from the global warming policies that they now endorse.

While relatively little is said in the Bible that directly addresses our use of natural resources, there is abundant advice on the importance of caring for other people. And that care does indeed depend, one way or another, upon our use of natural resources. But in order to make responsible policy decisions and avoid doing more harm than good to both the environment and humanity, it is imperative that we not repeat the mistakes of the past. I have come to the conclusion that these mistakes, like most policy mistakes, are usually the result of widespread misconceptions that exist in one particular domain: basic economics.

Chapter 6: It's Economics, Stupid

DESPITE THE PUBLICIZED ranting by some climate scientists, science by itself has nothing to say concerning what should be done about global warming. Science is policy-neutral and values-neutral. While the Union of Concerned Scientists is interested in what society should do about a wide variety of issues like global warming, their views should carry no more weight than, say, the Union of Concerned Movie Stars' policy position on global warming.

Nevertheless, we scientists are citizens, too. We have our own opinions—this book, for example—about what should be done to reduce any number of perceived threats to humanity and the environment.

I have noticed that a person's opinions on policy matters are almost always a result of their understanding of economics. We cannot meaningfully discuss what should be done about global warming, or any other environmental policy issue, without a good working knowledge of basic economics.

Unfortunately, while economic concepts are inseparable from the discussion of our response to the threat of global warming, economists' explanations of how economics works are typically so jargon-laden and obscure that my eyes glaze over just thinking about them. The good news is that the economic principles that are the most important to understand are relatively easy to grasp. Yet despite their simplicity, as well as the overwhelming historical evidence for their truth, many people still refuse to believe them. But you look like a reasonably intelligent person, so let's forge ahead.

The famous definition given to the term "economics" by the nineteenth-century economist Lionel Robbins is "the study of the use of scarce resources which have alternative uses." Another way of expressing this is that the practice of economics involves the exchange of our time and talents in ways that maximize how much stuff we all, collectively, get from the limited amount of stuff that can be produced. Putting it even more simply, economics involves people doing useful things for each other, hopefully in the most efficient manner possible.

Why should our policy response to global warming come down to economics? A progressive, environmentally conscious person might say, "Money, money, money ... all people are worried about is the bottom line, how much they can earn. The global environment is too important to reduce it to a matter of dollars and cents."

But what that person does not appreciate is that, except for social capital commodities such as love and friendship, *everything* comes down to money. Not money *per se*, but the relative value to humans of one thing versus another, which we quantify in units of money. Giving different things different monetary values is simply an easy way to quantify how important these things are to society. Humans cannot live without altering their environment to suit their needs, and smart economic decisions make sure that the needed natural resources are allocated (shared) in the most efficient ways.

Unless we understand basic economic principles, we cannot come to a responsible view of what should be done about global warming, or any other environmental issue that costs money to fix. So let's review some of the economic truths that I hold to be self-evident. While none of these concepts are new, they have been clarified and sharpened for me by two great economists: Thomas Sowell and Walter Williams. Here I present them in a way that has made the most sense to me over the years. Any loss of accuracy resulting from my own interpretations and examples is my fault alone.

There is no such thing as a free lunch.

Unlike some truisms, this one is always true. Think radio is free? You pay for it through higher prices for goods and services advertised on the radio. Free health care? Someone has to pay for it. For many Europeans, their "free" health care is paid for by charging $6 for a $1 gallon of gas. Do you believe that the salesman really is throwing something in for free when you buy something else? Try telling him you will take just the free item, thank you very much.

I assume that people want to keep eating. Clothes to wear? Some place to live? Transportation? Communication? Medicine when they get sick? X-boxes and iPods? A clean environment? All of these things (which I will interchangeably call "wealth" and "stuff"), require work and resources to produce.

And what about those who cannot provide these things through their own efforts—the poor, widowed, orphaned, chronically ill, and the elderly? Taking care of them requires even more wealth. And what about when a natural disaster strikes, and many people are unable to contribute to the economy anymore, but still require goods and services just to survive? Still more wealth.

Misconceptions about money can get in the way of our understanding of wealth. Money has no inherent value by itself. It is simply a mutually agreeable and ready form of exchange of individual units of wealth between people. Money allows the car manufacturer to sell his car to the baker without having to accept 21,000 loaves of bread in exchange. And the baker can sell a single loaf of bread to the car manufacturer without having to accept a car turn signal bulb in return.

How about the government printing more money? That sounds like an easy way to create more wealth! Unfortunately, printing more money creates no new wealth. The printing and spending of more money by the government has the same effect as raising taxes since, in effect, it lowers the value of all of the money that is already in circulation. There is more money chasing the same number of goods and services, which then causes prices to rise.

The practice of printing more money is a major source of inflation. Entire governments have collapsed for not grasping the fact that money is not wealth. In Germany after World War I, money was printed as fast as possible to pay for debts that resulted from the war. The inflation rate was astronomical. Money that employees made in the morning was almost worthless by the end of the day. There were not enough printing presses to print money fast enough. It reached the point where people had trouble just carrying the amounts of money needed to pay for daily necessities.

The only way to create wealth is for people to do useful things for each other. There is no free lunch, because it took time, resources, and human effort to make that lunch.

One opinion that is often voiced about global warming policy is that, given the global warming threat, we must *do something*. But "doing something" inevitably means devoting some portion of our wealth to attack that problem, which means that that portion of our wealth is no longer available to address other problems. Thus, the cost of doing something to fix one problem needs to be weighed against the use of those funds to address other issues. It's one of those "cost versus benefit" things you might have heard about.

Put another way, there are not unlimited financial resources to fix every problem that faces mankind and the environment. This is a specific example of the more general economic truth that people have an unlimited source of wants, but only a limited supply of goods and services. We all want more than we can provide for each other. This is what economists like to call "scarcity."

John Stossel, a consumer advocate reporter on ABC's *20/20*, years ago had a remarkable revelation. He finally understood some basic economic truths. In the first of a series of specials, he posed the question "Are We Scaring Ourselves to Death?" The title was meant to be literal, not figurative. His primary thesis was that, when we allow ourselves to be overly concerned about something that is a lesser threat, our spending of some portion of our wealth to solve that problem means that other, more pressing problems will likely get less money.

When the media decides what issues you should be informed about, they are unknowingly assuming a huge responsibility. The media shaping of public opinion on issues that the media decides are important can result in public policy changes that can literally kill people. Economists have been trying to tell us this for years, but we didn't understand them because they keep using words like "scarcity" and "marginal costs."

But shouldn't we be doing something about global warming as an insurance policy against future loss? Sure . . . if it makes economic sense. For instance, we buy homeowners' insurance to protect us against a loss that we cannot afford to replace. It makes sense to spend, say, $1,000 a year on homeowners' insurance that gives us 100 percent of the replacement value if a $200,000 house is destroyed by fire. But in the case of global warming insurance, most policies being promoted are like paying $10,000 a year on insurance that doesn't even begin to cover the replacement cost.

THE TOTAL AMOUNT OF WEALTH IS NOT CONSTANT.

Many people think that there is a constant amount of wealth, and all that matters is what you can do to grab a piece of the pie. In these folks' minds, there are the "haves" and "have-nots," and life is an unfair struggle for everyone to "get theirs." People who have this view are into class warfare; they are hateful toward, and jealous of, the rich. (By definition, the "rich" are people who make more money than you do.)

If it were true that the total amount of wealth is constant, how could we explain the higher standard of living that we have created over the years? Many years ago, only the wealthy among us could afford an automobile, a refrigerator, or even a microwave oven. Now, even most of those who live in poverty have these modern conveniences.

A constant amount of wealth necessarily implies that it really doesn't matter what we do when we work, since it obviously doesn't change the total amount of wealth anyway. If this were true, we could all have jobs as ditch-diggers or ditch-fillers. Half of us could dig holes in the ground all day long, and the other half could fill them up again. We could have everyone working diligently, with zero unemployment. But no wealth would be created. Where would we get food? Clean water? Housing? Clothes? Medical care? TV's? Computers? iPods? Automobiles, airplanes, and all other forms of transportation? Who would invent new and more efficient ways of providing these goods and services? Not only would no new wealth be created, but existing wealth would be destroyed, since it would be gradually used up or worn out.

It makes all the difference in the world *what* people do when they work, not just *that* they work. Whether you pay your neighbor $2 or $2,000 to dig a hole in your front yard, you still only have a hole to show for it. It is what we do for our money, and how efficiently we do it, that matters. The more efficient we are at providing the goods and services that other people need, the more wealth everyone will have.

PEOPLE GENERATE WEALTH, NOT THE GOVERNMENT.

There is a true story about someone in the audience of a TV talk show remarking, "the taxpayers shouldn't have to pay for this service, the government should!" From an economic point of view, the taxpayers are the government. For the most part, government does not generate any new wealth, except to the extent that it provides some services that everyone values (e.g., national defense), which we pay for with our tax dollars.

The government collects money from us through taxes, and

redistributes it to others based upon whatever priorities our elected representatives have decided are important. But the value of that "government money" comes from commerce carried out between people, not from some sort of governmental blessing it has been given.

In a healthy economy, it is the people who determine what and how much of different things they want, not some government bureaucrat. It is not just a theory, but a historically demonstrated fact that it is the people—not politicians—who are the most efficient at deciding what goods and services they need and want, and what the prices of those things should be. Every time a nation's leaders try to control prices or supply, the will of the people is thwarted. This is why political and economic freedom is so essential to the prosperity of a country, and why many countries, especially in Africa, are so poor.

Even though the government does not, in general, create wealth, it can certainly enable or discourage the generation of wealth by its citizens through its ability to pass laws and collect taxes. Any activity that is taxed more by the government will be avoided more by consumers and investors. Conversely, activities that are taxed less will be encouraged. Partly as a result of capital gains tax cuts, the economy in 2005 and 2006 was so healthy that hundreds of billions of dollars more in tax revenue was collected by the government than was expected. That is a lot of money in anyone's book.

Thus, more tax revenue can usually be collected by encouraging economic growth than by raising tax rates. Tax revenues are a "percent of the action," and anything that stimulates more action leads to more tax revenue. When politicians try to collect more tax revenue by increasing tax rates, they usually end up hurting the wealth creation process and, as a result, collect less tax revenue.

Free markets provide the most prosperity for a society.

As Adam Smith observed in his 1776 book, *An Inquiry into the Nature and Causes of the Wealth of Nations*, the selfishness of those who seek a profit in a free market economy simultaneously causes an "invisible hand" to reach out to help others. After all, the person who grows rich only does so through the willing participation of other people. Others either give some of their money to get valuable goods or services that the rich person and his business offers, or they work for the rich person to help produce those goods and services. Everyone benefits when these transfers are done on a mutually agreeable basis, as is done in a free market economy.

In a free market economy like that in the United States, it is the consumers (you and me) who make economic decisions. If something costs more than we think it is worth to us, then we will spend our money on something else that has greater value to us. On the selling side, if the seller is failing to sell enough of an item that is being passed over by consumers, he will be forced to lower the price to make the item more attractive compared to other items. If a profit cannot be made by the seller because the item, through inefficiencies, cost too much to produce, then someone else will figure out a better, less expensive way to manufacture and distribute the item. Free market economies ensure that a maximum amount of stuff is available at the lowest possible price.

Decades of misery persisted in the former Soviet Union because the government made supply decisions for the people. The huge inefficiencies that resulted were caused by the lack of the continuous feedback between producers and consumers that exists in a free market economy. Soviet economists were fully aware of the problem, but their government wasn't very tolerant of criticism of its economic policies.

Our high standard of living depends upon allowing prices to fluctuate with supply and demand. The alternative is for government to artificially fix both supply and prices, but doing so only makes the market less efficient at generating wealth.

Even in America we have dabbled in controlling prices, and always with bad results. Price controls sound like they would help to keep things cheap, but in the end they create shortages and subvert the overall wealth-building nature of free markets. Prices rising and falling are a sign that the system is working. In contrast, price controls benefit the few who can get there first, while the rest of us don't get any.

When there are shortages in energy, the free market system offers the best solution to the problem. If gasoline prices rise sharply because hurricanes destroy oil platforms and drilling rigs in the Gulf of Mexico, or because global demand for energy was greater than production, some people will naturally conserve more to avoid paying the higher price for gas.

For instance, in 2005 virtually all of our Gulf of Mexico petroleum production capacity was shut down by hurricane strikes, causing gas prices to rise sharply. But within a few months, the average price of a barrel of oil started falling again. That's because people were conserving in the face of higher prices at the gas pump. They stopped taking that extra five-mile trip to the grocery store just to buy the Q-Tips they forgot the last time they went shopping.

In contrast, the desire of California's politicians to fix energy prices at some "fair" level resulted in blackouts during the summer of 2000. Electricity shortages were made much more severe because rising prices were not allowed to force the consumer to reduce consumption. California utilities lost huge sums of money because they had to pay other states high prices to get extra electricity during periods of high demand, but they were not allowed to pass the price increases on to the consumer. In the end, Californians had to pay for the higher prices anyway, because the government had to bail out the utilities.

Despite the central role that profits play in enabling the prosperity of a free market economy, there seems to be widespread resentment of the rich. Does Bill Gates have more wealth than he "needs"? Maybe. But the promise of greater wealth is why people work so hard to find more efficient, and therefore lower cost, ways of providing goods and services. History has shown that if the profit motivation is removed, people tend to get lazy. Then everyone suffers.

If a few people get obscenely rich in the process, what do I care? The software products that Microsoft creates have made my job much easier and more productive. Those benefits, which are enjoyed by hundreds of millions of people around the world, are much more valuable to me than, say, my tiny share of Bill Gates' fortune should he decide to give it all back to the consumers.

Since Bill Gates cannot personally design, manufacture, and distribute all of these software products, he employs thousands of people who help him accomplish the task, who then share in the newly created wealth. Those Microsoft employees then exchange their money with all kinds of stores and merchants who, in turn, have their own employees.

The rich become rich only because consumers voluntarily give them money in exchange for the valuable goods and services they offer to society. The mere existence of the rich should remind us that the system still works, and that millions of people are benefiting from the innovative ideas of a creative few.

One of the biggest complaints about the free market system is that it is "unfair." There are wide disparities between rich and poor. But is it unfair that people be rewarded for their innovative work that leads to so much prosperity? For developing more efficient ways to meet the needs of their fellow man? For causing the creation of millions of jobs, thereby enabling many others to share in the newly created wealth? For creating the extra wealth that is needed to support (through charity) those who cannot support themselves?

Still, people continue to hold onto the mistaken view that by imposing "fairness" on the exchange of goods and services we can let everyone share equally in our wealth. While it does accomplish the equality part, it has the unintended consequence of making everyone equally miserable. There are no longer any incentives to maximize our creation of wealth, and so the economy suffers.

Despite these economic realities, the mainstream media continues to champion anyone who advocates such approaches to "fairness." World leaders like Cuba's Fidel Castro get glowing praise from people who themselves would not live in the economic conditions that have been imposed upon that country. There's a reason why Cubans continue to drown trying to escape to the United States. If you want to see why so many journalists are so clueless on economic issues, just look at the course requirements for a degree at any journalism school.

The poor in a free market system are typically richer than the poor anywhere else. Because of the great amount of wealth that a free market economy generates, there is plenty left over for charitable contributions to keep people afloat who cannot, for one reason or another, provide for themselves.

The personal charity of people after Hurricane Katrina led to the housing of hundreds of thousands of people who had lost their homes. The rest of the country, through taxes, will help to rebuild the hurricane-devastated portions of Louisiana, Mississippi, and Alabama. The only reason the United States can absorb such a catastrophe with so little damage to its economy is the economic wealth and infrastructure that has been built up over the years. And that wealth is only possible through free markets, allowing the *people* to decide what something is worth to them, rather than allowing government bureaucrats to decide.

In a socialist country there are few rewards for extra effort, for new ideas, or for improved efficiencies. Everyone gets the same, equal, and comparatively small share of the total amount of meager wealth that has been generated by the populace at the direction of the government. A country like Sweden has only been able to make socialism work for so long because they are not entirely socialist. They have kept free market principles in place, which has helped to generate sufficient wealth to support the outrageous level of taxation they now have. And recently, in 2006, the increasing desire to generate jobs over welfare handouts has led to the ousting of the Social Democrat government in Sweden.

Finally, the great wealth generated by the United States' free market economy has not just helped the United States. While many nations of the world seem to resent the wealth of the United States, the technological innovations and increases in efficiency we have spurred have benefited most of humanity. Other countries have reaped many of the benefits of America's inventions and manufacturing efficiencies.

This is one reason why we should not feel guilty that the United States happens to have the largest per capita carbon dioxide emissions of any country in the world. In a very real sense, we help feed the world, and we provide a vast variety of goods and services that have raised the living standards of the rest of humanity.

WEALTHIER IS HEALTHIER, SAFER, AND CLEANER.

Some Americans experience guilty feelings over our wealth as a nation, or over the fact that most folks don't have as much as them. Many environmentalists despise the West's modern way of life. Others feel that money has brought them more trouble than it is worth.

I believe that this collective angst is the result of many people simply having too much extra time on their hands—time which is only available because of the economic efficiencies we have created. There is a reason why only the wealthy nations of the world worry about the environment: we are the only ones with enough leisure time and wealth to afford that luxury.

Any guilt felt by the wealthy is needlessly self-imposed. If a rich person's wealth was created through the provision of goods and services that other people value, then you can bet that many more people have collectively benefited than the single, guilt-ridden rich person.

And what would the alternative be for people who think that wealth is bad? How would those same people like to deal with the angst of having half of their children not live to see their teenage years because of rampant disease, dirty water, poor nutrition, and food-borne illnesses from a lack of refrigeration? Until about a hundred years ago, this was the case for most of humanity. It is still the case for about one billion people today.

Or what about the angst of back-breaking labor just to make

sure your family has sufficient food, clothing, and shelter every day? It hasn't been that long ago that our level of wealth (say as measured by the Gross Domestic Product) began to skyrocket with the onset of the industrial revolution. Technological progress has given us new tools, new conveniences, new medicines, longer lives, and children who actually reach adulthood. Satellites and weather radars help to warn us of approaching hurricanes and tornadoes, saving many lives. In contrast, tropical cyclones in poverty-stricken Bangladesh as recently as the late twentieth-century routinely killed tens of thousands of people.

Do we really want to go back to the "good old days"? There are still people living today who lived through those days. If you talk to them, you will find that the good old days weren't so good after all. These people recount how miserable daily life was: dirty, smelly, full of illness, and dangerous. The widespread use of horses for transportation caused a continual stench, and many people were injured and killed by them.

The only reason everyone looked so fresh and happy on the TV series *Little House on the Prairie* was because those actors and actresses had just taken hot showers in their air-conditioned trailers after enjoying a catered lunch and were looking forward to their next big paycheck.

Bjorn Lomborg, in his excellent book *The Skeptical Environmentalist*, reviews in great detail, from the data made available in the U.N.'s own publications, how much better off both humanity and the environment are than they were in decades past. Most diseases that used to kill people before they reached adulthood have either been eradicated or now have cures. The production of wealth has led to the widespread availability of electricity, clean water, sanitation, and refrigeration to prevent food-borne diseases.

People who now yearn for the "good old days" do so from a position of wealth, health, comfort, and safety. Even though fossil fuels have been indispensable to the advancement of the human condition, some folks are seemingly now eager to dispose of what energized that progress. They are like spoiled children, biting the invisible hand that feeds them. Rather than celebrating all that is good, they focus on what is wrong, as if it were possible to achieve perfection, to build a utopia that, unfortunately, can only exist in people's imaginations.

Yet some still persist in the belief that such a utopia can be built, where wealth and equality of outcomes can coexist. They cannot put modern life in the historical context that gives it meaning and demonstrates its superiority over the other alternatives. They feel that things could be so much better, without realizing that we are now at a historic pinnacle of progress, health, happiness, and safety. They go through their lives wringing their hands about the inequities of life without realizing that most people, through their own decisions, have chosen their present circumstances.

They do not realize it, but most of them would not want to live in a world where everyone is forced to be equal.

From Economics to Environmentalism

Those who advocate the most impractical solutions to global warming seem to not understand how a free-market economy works—even though they participate in one. I'm sure most of them have pure motives, but to rephrase a famous saying, the road to environmental destruction is paved with good intentions.

It is imperative that we understand basic economic truths when considering policy approaches to fight global warming, or any other environmental problem. Other than social capital items like love and friendship, everything else that has value can be given a price. Does a forest of trees have more value as pulpwood and lumber, or as a place for folks to just enjoy nature? That decision comes down to how much people are willing to pay for one use versus another.

Let's say a person, we'll call him Jeremiah Johnson, owns forty acres of pristine wilderness, and that is literally all he owns. Jeremiah has to figure out what he will do to get enough food, clothing, and shelter to live. He might start by trying to do everything himself. He might fashion some sort of clothes out of grasses. He might eat fruit and berries that he gathers in the forest, and plant some sort of crops. Jeremiah could start building himself a log cabin for shelter. He would just have to hope, for the time being, that he did not develop an illness or have a bad accident that would require medical attention.

Then, one day, someone visits Jeremiah and offers him a trade. If Jeremiah will give the visitor ten of his forty acres, he will receive in exchange food, clothing, and a house that will be built for him on his remaining thirty acres. For somebody in that situation, it would be difficult to pass up such an offer. If you have ever been primitive camping, you know what I mean.

Jeremiah has thus given a monetary value to his land. Monetary value is simply an expression of how valuable something is to people *compared to other things*. I, like most others, believe that the United States should set aside wilderness and parks, and keep them free from development. But that is because some kinds of land are especially valuable to the public. It is part of our collective national wealth. The decision to preserve that land is, ultimately, an economic one.

And it is its monetary value that ends up actually protecting privately owned land. In order to maximize the resources that the owner can harvest from his land, the landowner is naturally incentivized to sustain its value. The land will be of no value to anyone if it is destroyed. In order to maximize and sustain his profits, the landowner must care for the land in a responsible manner.

Is "Big Oil" the Enemy?

As I have mentioned previously, if cost was no concern and we had easy alternatives for fossil fuels, then switching to those alternatives would be a no-brainer. But this is not the case. There are, as yet, no practical large-scale alternatives to fossil fuels.

Still, there seems to be a widespread perception that energy companies are somehow conspiring to keep lower cost and environmentally friendly energy alternatives from the masses. If only we would just switch to solar and wind power, our global warming problems would be over, these folks reason. The energy companies are *trying* to poison the environment, just to spite environmentalists! *Mwa-ha-ha!!*

> *Cigar-chomping oil company executive #1:* Hey, we haven't heard much lately from that environmental group. What's their name?... oh, yeah, "Earth First, Humans Last."
>
> *Cigar-chomping oil company executive #2:* You know, you're right. How about we cause an oil spill from a pipeline somewhere? That'll get 'em riled up!

Do we really think that energy companies enjoy being hounded by environmentalists and the public? If the price of gas goes up, they are accused of price gouging. If it goes down, they are accused of making it too easy to pollute because of cheap gas, or of giving in to political pressure from the White House. As I said earlier, people's understanding of basic economics also affects their political views as well.

Claims of price gouging ignore the fact that, due to real fluctuations (and forecast future fluctuations) in global supply and demand of petroleum, the price of a gallon of gasoline at the pump can fluctuate by as much as 50 percent in a single year for purely market-based reasons. Sure, when the price goes up, the oil companies can reap massive profits. Any industry which provides goods that everyone needs will make lots of money when reduced supply (or even the threat of reduced supply) causes prices to rise. After all, as mentioned earlier, the hope for large profits is what makes a free market economy so efficient at raising the standard of living for everyone.

But the petroleum industry has down years, too. They need funds to repair oil rigs, platforms, and refineries after a Category 4 hurricane rolls through the Gulf of Mexico and comes ashore. Would those who advocate a public redistribution of oil industry profits during time of high prices be willing to bail out those same companies out when disaster strikes? I don't think so.

Also, we should also keep in mind that those cigar-chomping executives don't own the oil companies. Public investors do. And for the most part, it is those investors who experience these gains and losses experienced by the industry, not the employees of the company.

In a free market economy like ours, the cost of alternative fuels is automatically taken into account, and determines to what extent they will be used as an energy source compared to other energy sources like coal and petroleum. A petroleum company employee once told me, if their company can find an economical source of energy other than fossil fuels, they will promote it. They are in the business of making money by supplying energy to consumers, and it doesn't matter to them what the source of that energy is.

If wind energy is indeed less expensive than coal-fired power plants in some portion of the country, then its use will gradually grow. If solar power is economically competitive, its use will also grow. But sometimes the government subsidizes uneconomical alternative technologies. In a free market economy, such artificial support for non-competitive technologies does not spur long-term investment in those technologies. Investors know that as soon as the artificial supports go away, any profit potential disappears. You can tell how much potential a future energy technology has just by how much private money is invested in its development.

Fortunately, the free market economy takes care of these issues automatically. If some future energy technology is indeed promising, then investors will support it and speed its development and deployment. The government doesn't need to do anything for this to happen, except to stay out of the way.

The Economics of Pollution

I am continually perplexed by environmentalists' attitudes toward factories and big business. The greatest number of products, with the least amount of wasted resources, energy, and pollution, results from centralizing the making (mass production) of those products in large manufacturing facilities.

Despite these efficiencies, many people complain, "Isn't it awful, all of the pollution that factory is producing!" But, generally speaking, that factory is producing much less pollution and fewer wasted resources than if every family or every city was responsible for making those same products for themselves. All of the leftovers from the production process just happen to be concentrated in one place, giving us a false picture of the net effect of mass production on the environment.

Certainly we need to keep this concentrated, centrally produced pollution from causing undue harm to the local environment or to human health, but the total amount of that pollution has already been minimized to a large extent through the economies of scale. Further reductions in pollution beyond some reasonable level are an economic decision. If such further reductions are mandated by the government, it costs some portion of our wealth to accomplish those reductions. How much more are we willing to pay for goods and services to achieve ever-increasing levels of cleanliness?

When the EPA mandates the reduction of some types of pollution to lower and lower levels, they seem to be unconcerned about what the cost to society will be. I was astounded when, at a recent air pollution control conference, an EPA official actually told the audience that "we can't stop pushing" for a cleaner and cleaner environment. This is a dangerous position to take, since there comes a point where the benefits of "cleaner" become too meager, and the costs too high. The result is that some of the limited wealth available to attack all of society's problems is no longer available to address some other problem of greater importance.

As a result, government regulators working for the EPA generally don't have to worry about the costs imposed upon society of reducing pollution by ever-increasing amounts. First they will mandate a 90 percent reduction in some pollutant. Then 99 percent. Then 99.9 percent. It never stops, because that's their job: to keep reducing pollution.

But no matter how hard we try, it is physically impossible to eliminate pollution. Even if it was possible, no one would want to pay the enormous cost of accomplishing it. Take your home as an example. You do not allow filth to build up to the point of being a health hazard. But, unless you happen to be obsessive-compulsive, neither do you spend inordinate amounts of time and energy making sure every surface is germ-free. Just as it costs something to hire a janitor or housekeeper, reducing pollution also costs something. And the cleaner we try to make things, the more it costs.

I have demonstrated the cost versus benefit relationship for pollution control efforts in the accompanying graph. (Scientists love graphs.) As environmentalists push for a cleaner and cleaner environment, the costs skyrocket. Since society has unlimited wants, but only limited financial resources, these costs must be balanced against the costs of addressing other problems facing society. As the graph demonstrates, the cost of a totally clean environment can be unacceptably high.

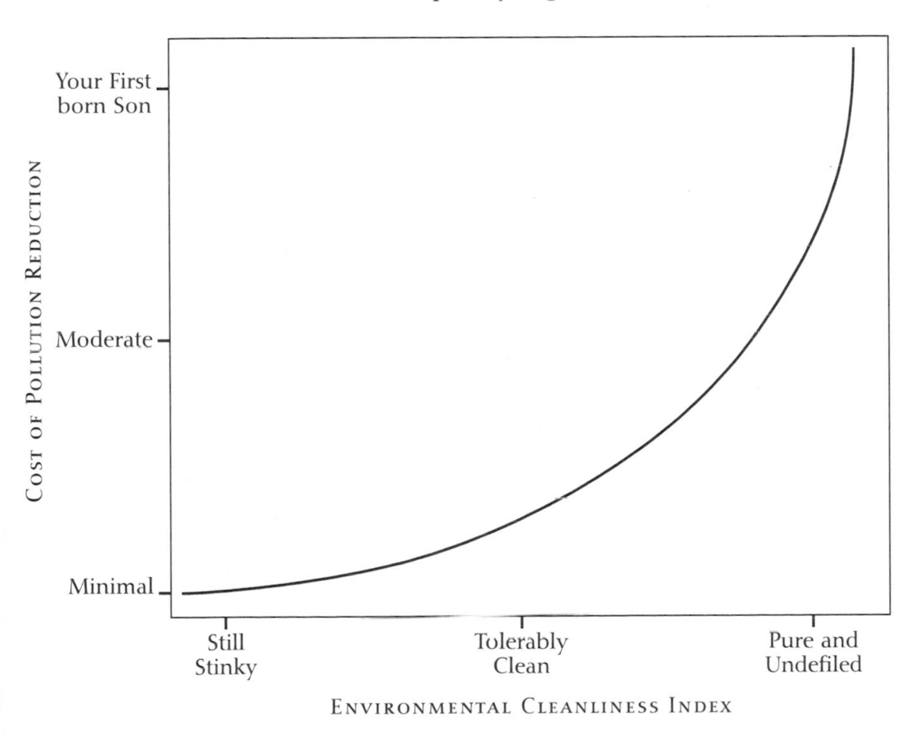

When the EPA demands an industry to "clean up," do you really think they are going to absorb all of the cost of doing so? As a general rule, competition in the free market has already minimized their profit margin, which historically averages around 10 percent. The citizens who own stock in the company are the ones who share in those profits. Additional costs for pollution control in industry will simply be passed on, first to consumers, and then to shareholders in the form of lower profits.

As a result, when the public demands that some industry clean up its act, they are implicitly agreeing to pay the cost for that cleanup.

All too often, industrial pollution becomes a political football that environmentalists and politicians use to pander to public emotions. Take the example of mercury pollution from coal-fired power plants. This source of pollution has remained unregulated since it started in the late 1800s. There is no empirical evidence of negative health consequences of mercury pollution from these sources. Japanese cancer rates are much lower than those in the United States, even though the Japanese have much higher mercury concentrations in their bodies from their greater consumption of fish. Dentists have been handling mercury for years with no demonstrated ill-health effects.

Nevertheless, since mercury is perceived to be public health hazard, the Bush administration proposed a cap and trade regulatory mechanism to reduce this source of mercury pollution by 29 percent by 2010, ramping up to 79 percent by 2018. The cost to the electricity industry (which means the cost to you and me) was estimated to be about $2 billion. But then, despite that fact that there had been no mercury pollution controls up until that point, environmentalists suddenly demanded that the pollution be reduced by a much greater amount, and on a much faster schedule. It was estimated that this would have cost over $300 billion to accomplish. Some critics even suggested that President Bush's "weak" plan was evidence of some stealth conspiracy to hurt the nation's children.

Well, I have decided that even the environmentalists' plan is not stringent enough. The electricity industry clearly needs to stop all mercury emissions within one month. These so-called "environmentalists" obviously don't think our children's health is important enough to eliminate this dangerous pollutant right now. They must be out to hurt our kids, since their proposed emissions reductions do not eliminate the threat, and take so long to implement.

I think you see my point. It is vitally important that environmental regulations be fashioned to maximize benefits and minimize costs. This also means that the public should not be fooled by political rhetoric which claims some politician "wants" to pollute the environment, or is "against environmental regulations." Every decision we make in our daily lives balances costs versus benefits, and the same balance needs to be applied to demands by some for an increasingly clean environment.

In summary, the policy response to global warming, just like any other environmental challenge, always involves economics. This chapter is meant to provide a better basis for you to judge for yourself whether proposed solutions to global warming (or any other environmental ill) make sense or not. We cannot simply mandate reductions in carbon dioxide emissions without also determining what negative consequences will likely result from such a policy.

In the coming years, there will be increasing pressure to "do something" about global warming. As we shall see, the risks of most currently proposed pollution reduction policies far outweigh the benefits. So, when politicians start claiming to be "doing something," you need to ask two questions: "How much will it cost us?" and, "How much will it help?"

Since the proponents of dangerous and overly restrictive policies typically have political motivations, let's take a look at some of the players in the global warming policy game. We will see that none of them can claim to be unbiased defenders of the moral high ground.

The Environmental Protection Agency cranks it up a notch.

Chapter 7: The Politics of Climate Change

THE COSTS AND consequences of "doing something" substantial about global warming in the near future are staggering. All of humanity needs a source of abundant and inexpensive energy to thrive, and no matter what energy policy changes are implemented by governments, there will be big winners and losers, financially and politically.

If governments (rather than people in a free market) can control what kinds and amounts of energy are acceptable, they will have vastly increased their control over their citizenry. Since global warming respects no international boundaries, the United Nations' dream of global governance is now closer than ever.

But while global warming might be portrayed as an enemy against which politicians from around the world can unite, it is more accurate to say that global warming is an opportunity to accomplish selfish goals that might never have been achieved through legitimate means.

THE RESEARCH PLAYERS

Let's start with the little guys—like me. The scientific research establishment is widely assumed to be the unbiased purveyors of global warming knowledge, trying to fight back political pressure that would keep them from revealing to the world how serious the threat of global warming really is. But trust me, scientists are not unbiased, and you can rest assured that you have already heard about every possible catastrophic scenario that our fertile scientific imaginations can dream up.

With very few exceptions, we climate scientists are funded by the federal government. We write research proposals to NASA, NOAA, NSF, or the Department of Energy. It is you, the taxpayer, who are footing the bill for our research. I assume that you would like to see your money spent in such a way that is not biased toward any particular political persuasion. Unfortunately, both the scientists and managers within the funding agencies tend to have political biases and financial motivations which then influence how they approach the global warming problem and its solutions.

In my discussions with various climate researchers and funding managers over the years (typically over a beer or three), I have found that many of them are closet socialists. Most would probably dispute this, or don't even realize it, but the politics they support are pretty close to being socialistic. At the very least, they believe that we must "do something" about global warming, and most of them specifically support the Kyoto Protocol approach. (The Kyoto treaty will be addressed in more detail in the next chapter.) Some have even admitted that they would hold the same opinion even if global warming ends up not being a threat to civilization, which is a sure sign that something more than science is influencing their opinions.

Like so many other affluent westerners, most scientists and their governmental funding managers share a worldview in which mankind is ruining the Earth. This paradigm influences the kinds of research programs that are formulated, the proposals that scientists write to receive funding from those programs, and even how the research is carried out. While I am not accusing anyone of scientific misconduct, I am saying that it is difficult to make global warming research totally independent of political beliefs and worldviews.

Remember, we scientists do not provide goods or services that are useful to the public, at least not in a direct way. We don't have to "earn" a living in the normal sense of the word. As I explained in the last chapter, this makes scientists a little disconnected from what makes a free market economy work. It isn't our job to create wealth—it is to spend it. Besides, our PhDs are proof that we are above any provincial, mundane tasks like providing goods and services that folks need on a daily basis. We have a higher calling. We deal in the Discovery of Knowledge.

What I find amusing about many scientists' views on global warming policy is that they are contrary to the process by which the scientists reached their comfortable position in life in the first place. A democratic form of government and free market economy are what allow sufficient wealth to be generated to provide these scientists the luxury of pursuing their research interests. The United States funds most of the climate research being performed in the world, which has now amounted to many billions of dollars.

Yet the majority of these researchers are totally unaware that they are pawns in a political power game that cares little about whether global warming is a threat or not. Even in a subconscious way, scientists know that by playing this game in which mankind is the enemy of the environment, they can propose to a government agency to gain funding. And if scientists can help the government administrators build their global warming research programs still further, it is more likely that they will continue to get funding.

And, as a side benefit, we get to Save the Earth, too! What a great gig this is!

Global warming researchers were funded to study evidence for manmade global warming, not against it. Thus, their research results have a built-in bias that supports the theory of manmade global warming. Their published research is then inevitably colored by these biases.

As a result, we have papers published in the scientific literature that claim that this or that human influence is destabilizing the climate, which is then pointed to by the funding agencies as evidence that more money is needed to study the problem. These are not totally objective conclusions, as they have been biased by the way research questions have been posed and investigated.

I am not suggesting that there shouldn't be global warming research programs, nor am I suggesting that most of these researchers only purport to believe in manmade global warming just to get funding. As scientists, we all will agree that global warming could, at least theoretically, be a serious problem. Instead, I am merely pointing out that the research scientists, as well as the government managers that support them, cannot be considered to be unbiased. We all have a dog in this hunt.

I have often wondered if it would have been more fruitful for the federal government to request climate research proposals that would fall into one of two groups, with a goal of funding an equal number in each category. While one group would investigate evidence for climate destabilizing mechanisms, the other would look for evidence for climate stabilizing mechanisms. I don't know whether such an approach would work, but it seems like it would help to diffuse the bias inherent in global warming research today where researchers are falling all over themselves trying to discover some new negative consequence of global warming.

Through the peer review process, scientists help government managers decide which research proposals to fund. This is good, since the managers seldom have sufficiently detailed knowledge to make decisions about what scientists have proposed to do. But since the scientists involved in the review process are themselves chosen from the same pool of researchers, there is some level of professional incest that exists.

Compounding the problem is that fact that research disciplines have become so specialized that there might be only a half dozen people in the world qualified to review each other's proposals. This process further entrenches specific political and financial biases that already exist among scientists and managers.

If you still doubt whether there is inherent bias among the global warming pessimists, consider this. Imagine if the global warming threat were to disappear—for instance, some scientist convincingly demonstrates that we really do not have anything to worry about. Do we really believe that the environmentalists, scientists, and funding managers would breathe a collective sigh of relief and say, "What wonderful news for humanity! Now we don't have to worry about this problem any longer!"? I don't think so. Entire careers and scientific reputations which now depend upon global warming continuing to be a serious threat would simply end.

Government managers of climate research programs have to play up the threats of global warming in testimony to Congress in order to get a maximum level of funding for their programs. This is their job. Part of the reason for this is survival, since other agencies seeking funds are doing the same thing. And, admittedly, the global warming horror scenarios they paint for Congress might well materialize in the future. But when the whole research program is centered around, and even encouraging, the finding of evidence for an unstable climate system, you can bet the results will be biased in that direction.

When NASA was selling its *Mission to Planet Earth* to Congress, some legislators were honestly expecting that we would have global warming answers soon after launch of the new satellites that NASA was developing. Surprise! Nearly ten years later, we are still trying to figure out from all this satellite data how sensitive the climate system is to manmade greenhouse gases.

In my experience, government managers shy away from exerting direct pressure to come up with specific scientific results, or to change a scientist's testimony to congress. But more subtle pressures do exist. In my congressional testimony as a NASA scientist, I was reminded to limit my testimony to my area of expertise, and not to be drawn into policy discussions, in which I was not an expert. At that time, this meant I could talk about our satellite-based global temperature measurements, and nothing else.

I knew that my agency's research program could be hurt if I expressed doubts about the "manmade" part of global warming theory, and so I accepted the "advice" like a good employee should. It is almost inevitable, however, that during congressional testimony a senator will ask, "What would you do about global warming policy if you were me?" And when that finally happened to me, I so artfully dodged the question that members of the committee laughed. They said I sounded like a politician. Ouch.

In contrast, other NASA employees that were more in line with the status quo in their global warming views didn't seem to be dissuaded from offering more dramatic opinions in their testimony. For instance, Dr. James Hansen of NASA in 2006 made a pretty big deal about being pressured by the Administration regarding his interactions with the media. The public affairs office at NASA Headquarters started pressuring Hansen regarding what Hansen wanted to say, and the administration had some concern over whether some of Hansen's conclusions about global warming could be supported by the evidence.

What the public wasn't told during all of this was that NASA public affairs always wants to be kept in the loop regarding NASA employees' interactions with the media. Understandably, NASA managers do not like to be blind-sided by reading what their employees have told reporters in the morning newspaper. It's part of the rules which I accepted as a NASA employee, and I tried to abide by those rules.

Regarding any "meddling" by the Administration in NASA's business, NASA is an independent agency within the government's Executive Branch, and so NASA and its employees answer to the president. He's the boss. NASA employment is a privilege, not a right, and NASA has historically liked to present a unified message for public consumption.

But as our godfather of global warming research and public

awareness, Jim Hansen had more political capital to spend than I did, and complained to the media. It is my opinion that Dr. Hansen had become accustomed to saying whatever he wanted, whenever he wanted, to whomever he wanted, on the science and policy of global warming. It sounds to me like the Administration might have simply asked NASA to start enforcing its own rules and Hansen balked.

Who knows? If I felt like I was on a mission to Save the Earth and was in his shoes, I might well have done the same thing.

But instead, in contrast to Dr. Hansen, I finally tired of the restrictions on what I felt I could and couldn't say to the media, and I voluntarily resigned from NASA in 2001. I didn't make any big media splash about the issue, and harbored no resentment over the matter.

As the U.S. government's leading global warming researcher, Hansen's job is probably secure no matter what he says or does. On the positive side, the whole episode has probably made NASA management more tolerant of diverse views being presented to the public by NASA's employees. In the case of scientific research, I'm afraid I don't see how a unified scientific message can be presented to the public by any governmental agency unless some scientists are, in effect, muzzled. Scientific research inevitably leads to a variety of opinions. The only way to avoid more than one scientific opinion is to fund only one scientist.

Is important scientific information being withheld from the public on the subject of global warming? No. One way or another, every possible global warming horror scenario has been already beaten to death by the media. You haven't missed anything. Other than the moon landings, Area 51, and NASA's weather control machine, we government scientists have no secrets to keep.

Of much more concern to me are the politicians with agendas that will now start making a big deal about perceived muzzling of scientists, further wasting time with political spin to try to discredit other politicians.

Politicians, Congress, and the EPA

Members of Congress fall into one of two camps on global warming. They either already have an established opinion on the subject and are looking for scientists that will tell them what they want to hear, or they genuinely want to understand a range of views so that they can make an informed decision.

Ha-ha! Okay, I was just kidding about that second group.

In any case, most politicians recognize the potentially huge impact that policy changes will have on the economy. That is why the U.S. Senate unanimously passed the Byrd-Hagel Resolution in 1997, 95-0, stating that the sense of the Senate was that it would not ratify the U.N.'s Kyoto treaty to reduce greenhouse gases. Since it did not include any restrictions on the developing countries like China and India, companies in the U.S. could just move to other countries with fewer environmental restrictions and pollute even more.

So, while some politicians would have you believe that it was the Bush Administration that stood in the way of adoption of the Kyoto treaty, they are simply letting someone else take the heat for a position that most of them still take as well.

In all fairness, Congress is stuck between a rock and a hard place when it comes to global warming policy. There is a constant roar of voices from environmentalists, and even much of the public, to "do something," and yet public surveys show that people don't want to "do something" if it is going to cost them very much. The political commentator and funnyman Bill Maher made an excellent point when he asked how many of us would give up our TV remotes if that was all it took to avoid global warming.

The business community reminds Congress that "doing something" will hurt business—which, as the last chapter on economics demonstrates, means all of us. We all will suffer economically if we are not smart about global warming legislation. As long as we consumers want to continue to buy our stuff at the lowest prices, we *are* big business. If factories and electric utilities are forced to spend money to reduce greenhouse gas emissions, do you really think they are going to take it out of their hide? We are the ones who will pay for it.

While some politicians do indeed approach the global warming policy problem from a very pragmatic point of view, there is one politician for whom the global warming issue is spiritual and personal. His rhetoric sets him apart from other politicians because of the passion he has for Saving the Earth. He is the Former Next President of the United States, Al Gore, Jr.

If James Hansen is the scientific godfather of modern global warming research, F. N. POTUS is the political godfather of modern global warming policy. Mr. Gore deserves the credit for helping to bring the potential threat of global warming into the public consciousness in 1988 while he was a U.S. Senator. During a hearing that was allegedly scheduled on a day that was forecast to be unusually hot, Senator Gore had Hansen testify on the possible role that manmade global warming might have had in the drought that the Great Plains was experiencing that year.

This led to Dr. Hansen's shocking testimony that that he was 99 percent certain that some part of the drought was probably caused, to some extent, by the likely influence of manmade greenhouse gases ... maybe. Dr. Hansen thus became the first successful purveyor of scientific obfuscatory exaggeration, in which one can state something in carefully phrased, yet biased terms so as to cause a maximum amount of alarm, without being factually incorrect.

It appears that Mr. Gore has chosen to ignore all of the inconvenient truths that do not support the catastrophic view of global warming. He has surrounded himself with only those scientists who have bought into the present culture of global warming alarmism. Most reporters similarly get much of their juicy input from these global warming pessimists.

I do consider Mr. Gore to be a relatively science-savvy person. But his 1990 book, *Earth in the Balance: Ecology and the Human Spirit,* makes it clear that the issue is also a profoundly spiritual one for him. He has claimed that a wide variety of human activities, such as driving cars, should be done away with. As I addressed in

Chapter 5 (The Scientists' Faith, The Environmentalists' Religion), such spiritually-based motivations for changes in public policy come very close to being a state-supported Pagan religion.

The year after Hansen first testified I had the honor of being asked by Senator Gore to testify on our new satellite measurements of global temperature variations over the previous ten years. It being my first experience at providing congressional testimony, I marveled at the beauty of the hearing room. Wow ... high ceilings ... cool molding ... this place looks old.

A C-SPAN TV camera was being set up to record the hearing. As the starting time approached, I realized that Senator Gore, the committee chairman, would be presiding over this one alone. None of the other members of the committee showed up. I asked a staffer, "Doesn't this look bad, him being the only member here?" The answer was, "No, it's better this way ... he gets all the camera time."

The first scientist to testify in the hearing was Phil Jones, a British scientist who is best known for his development of a global surface thermometer temperature record extending back to the 1800s. The overhead projector that he would be using was seriously out of focus, and the focusing dial was obviously not fixing the problem. So, while I stood at the projector trying to remember my optical physics, Senator Gore was looking at me (thinking I was Phil Jones) and recounting the quality time "we" had spent together when he visited England. While this seems kind of humorous now, Mr. Gore probably has a better memory than I do for faces—I have only two children, and I still get them mixed up.

Mr. Gore remained dedicated to the issue of global warming during his public service. A few years later, there was a major weather event in the United States that Vice President Gore flew to in order to examine the damage and console the victims. A high-level weather expert, whom I'll call "Dr. Expert," was also on that flight. Vice President Gore asked Dr. Expert whether the severe weather event could have been the result of global warming. Dr. Expert said, well, probably not. The Bermuda high pressure area had stalled, leading to a persistent flow of moist air, blah, blah, blah.

Then, a few minutes later the V.P. was overheard telling President Clinton's diminutive female assistant, "Hey, Dr. Expert said this might have been caused by global warming!" The assistant looked up at the Vice President, and responded with something to the effect of, "Al, these people we are going to visit are suffering. The President doesn't want to hear about your global warming crap."

You can't say that Gore isn't passionate about the global warming issue.

I have to confess to not acting in a very professional manner at times when dealing with politicians. Maybe I'm just trying to see if they have any sense of humor. I was giving a talk at the National Press Club on some global warming mumbo jumbo, and the speaker just before me was Senator Chuck Hagel. Before the event started, I was shooting the breeze with the Senator. (His chief staffer later told me we can't call him "Chuck," only "Senator Hagel"). The Senator knew I was a NASA employee at the time, and we were discussing the landing of the first successful Mars rover, Sojourner, on July 4, 1997. I then "let it slip" that, "it sure looked realistic ... you would never know we put that whole thing together on a Hollywood sound stage." Senator Hagel looked genuinely concerned. I'm pretty sure he wasn't amused.

I also tried that line on Art Bell's popular Coast-to-Coast AM radio show, which routinely addresses government conspiracies and alien visitations. My comment was followed by a few seconds of dead silence. Once again, no sense of humor.

But seriously, folks .... the United States Congress has a history of making knee-jerk policy decisions based upon the testimony of only a few alarmist "experts"—two of whom are real experts, and the third being an actor who played an expert in a popular movie on the subject. Fortunately, Congress is gradually recognizing that there must be greater scrutiny of scientific findings that end up influencing public policy. Congress is tired of making bad policy decisions in response to a single scientific study, only to find later that the results of the study were disproved. An even bigger problem is Congress passing feel-good legislation that has short-term benefits for the legislators, but long-term negative consequences for the rest of the country.

You are probably not aware of how flimsy the science was that led to acid rain legislation. The National Acid Precipitation Assessment Project (NAPAP) was a ten-year research effort to determine the causes and effects of acid rain. In 1990, after ten years of study, it was concluded that prior fears of widespread acid rain damage from industrial pollution to lakes and forests were largely unfounded. Only one species of tree at high elevation was noticeably affected, and most acidity in lakes was traced to natural causes.

Nevertheless, the regulatory groundwork had been laid at the Environmental Protection Agency, careers established, promises made, and so expensive acid precipitation legislation was passed. At the very least, we can say that the acid rain threat was greatly overblown, yet most of our citizens still do not realize this. Fortunately, our country produces enough wealth to be able to afford the extra cost of partly cleaning up smokestack emissions to abide by Clean Air Act regulations. Cleaning up carbon dioxide emissions is another matter entirely.

The EPA deserves special mention when it comes to the politics of climate change. I remind you that government agencies have two central goals. The first is to forever perpetuate their own existence. Once these agencies are created, it seems they can never be destroyed. While the President of the United States has only temporary job security, it is almost impossible to get rid of a rank and file government employee. The second goal of a government agency is to spend as much of your money as they can get their hands on. That is their job.

The mindset that pervades federal agencies is usually diametrically opposed to the basic economic truths of environmental policy that I reviewed in the last chapter. The EPA is in a never-ending quest for more and more stringent pollution regulations. A country can't have too many environmental laws, you know. Just ask the environmentalists. Some environmentalists seem to live in a dream world where pollution is optional. They don't realize that it is impossible to not pollute. They won't be happy until that last 0.00001 percent of the pollution has been eliminated, no matter what the cost. And what happens if anyone tries to fight overly expensive and restrictive environmental regulations? They are accused of being enemies of the environment, in the pockets of Big Business and Big Oil, or out to destroy the health of our children.

Politically, the EPA depends upon activist environmentalists. Call them the EPA's cheerleaders. With the environmentalists' help, the EPA is the altruistic defender of our fundamental right to clean air, clean water, and clean dirt. And if the EPA is our champion, Big Business must be our enemy.

THE ENEMY: BIG BUSINESS

Politicians pander to the resentment that the public has toward "big business." As I have already mentioned, you and I are big business. From a basic economics standpoint, we consumers willingly give our money to corporations in exchange for goods and services that we value more than the other stuff that money could have bought for us. If a corporation, its executives, and its investors become obscenely rich in the process, it is only because we have "voted" with our money to make them that way.

Even though our high standard of living actually depends upon allowing people the opportunity to become rich, it seems like we can't help being resentful toward them when they succeed. We like our high standard of living, but we don't want others who have spearheaded that success to profit from it. Jealousy is an ugly thing.

I have come to believe that political pandering to class envy is the motivating force behind many proposed policy solutions to the global warming problem. People have a basic desire to see everyone equally sharing in the abundance of a society's productivity. While this is a laudable goal, it is impossible to achieve. As history has clearly taught us, maximum economic efficiency at producing wealth is only achieved when we are willing to reward the talents and creativity of the relative few among us who develop those efficiencies.

You can have equality of outcomes, or abundance, but not both at the same time.

If the profit incentive is removed, competition disappears, and then you can say goodbye to much of our prosperity.

A lot of politicians, like many citizens, hold these mistaken views of how wealth is built in a free market economy. This is especially true of career politicians—those who have never been part of the wealth-building process as, say, the owner or CEO of a company. If I were King of the United States, I would decree not only term limits on elected officials, but also a requirement that they have some prior experience actually doing something economically useful before running for office

As a result of common misunderstandings about how a free market economy works, we see congressional investigations into the "windfall profits" of oil companies when the price of gasoline rises abruptly. In the case of petroleum, it is global supply and demand that determines the price of gasoline (before taxes, anyway), not some cigar-chomping oil executive. There are multiple oil companies competing for your business, and competition keeps prices as low as possible given the existing supply and demand. But I suppose it is easier to just hate the rich than it is to face economic realities.

Price fixing through collusion between companies is extremely rare in a free market economy, simply because competition keeps it from happening. A price-fixing conspiracy would have to be kept secret across an entire industry, and then any competition that arises to offer lower prices would have to be secretly thwarted. In the case of a global commodity like petroleum, the conspiracy would have to be international. You probably can't get executives from five different oil companies to conspire to have lunch together, let alone pull off a global price-fixing conspiracy like that.

As long as we hold misguided views about the role of big business in the prosperity of the country, we will continue to waste time chasing our tails with regulation and taxation experiments.

Politicians and environmentalists will continue to paint business as the enemy in their efforts to gain your support for their cause. Many politicians are more than happy to take advantage of widespread misunderstandings on issues related to global warming. They increasingly pander to public perceptions regarding wealth, big business, and pollution. This trend cannot be sustained without seriously hurting the economy.

And there is a very good reason why environmentalists should also be concerned about hurting the economy. When economic hard times hit, taxpayers will start to jettison their concerns about superfluous issues—like environmentalism.

The global warming issue now provides politicians with the ultimate weapon to push for policies that are anti-freedom and anti-prosperity. If big business can be painted as the polluters, instead of you and me, politicians will continue to accumulate power at the expense of our prosperity and freedom.

As we will see, all proposed policies to fight global warming will have no measurable effect on future global temperatures anyway, and will definitely hurt the economy (the poor being the most vulnerable). In fact, we will see that economically damaging policies could actually delay the development of real solutions to the global warming problem.

Western powers make their contribution to resolving environmental issues

Chapter 8: Dumb Global Warming Solutions

WHEN FACED WITH a threat like global warming, it is only natural for people to want to "do something" about it. The trouble is, it is not obvious what can be done that will make much difference in the foreseeable future. Mankind needs an abundant source of inexpensive energy in order to prosper, and for now fossil fuels fit the bill. Any alternative energy sources currently proposed to reduce manmade global warming will have little impact in the next twenty years or so, no matter what you believe about future levels of warming.

I'll admit to being conflicted on the subject of renewable sources of energy like wind and solar. While these do have their place in the energy mix, their ability to help the global warming problem is pretty limited. I routinely encounter people who argue that renewable sources of energy can "fix" global warming. But the global demand for energy is so large that renewables will probably never be able to substantially contribute to our needs.

Everyone would love to be able to say they use a renewable source of energy—for instance solar or wind—but when you hand them the bill, or tell them a wind farm is being constructed next door, they start having second thoughts.

There are no economical alternatives to fossil fuels on a large enough scale, at least not yet. And given that we will continue to rely mostly on fossil fuels for the foreseeable future, environmental organizations and some politicians would have you believe that we must attack that problem through punitive measures. For instance, the government can mandate limits (caps) on carbon emissions, invoke a carbon tax, or make the industrialized countries of the world buy the "right to pollute" (carbon credits) from poor countries.

I have come to the conclusion that politicians who advocate such "solutions" are either incapable of critical thought, or have underlying political or financial motives. It almost seems like they want to pander to public sentiment, spread socialism, or destroy our modern way of life. Maybe all three.

As explained in Chapter 6, any global warming solution needs to be seriously examined in terms of costs versus benefits. Some politicians advocate policies that give government more control, rather than support efforts that will actually fix the problem. Despite what their proponents claim, though, most policies now being discussed are basically all pain for no gain.

There is a real danger in spending inordinate amounts of time arguing over policies that have no hope of fixing the problem. I see commercials advocate, and hear politicians talk about, taking steps that are little more that just exercises in making us feel better about ourselves. While one might think that such efforts are harmless enough even if they do fail, they represent a waste of time and resources that could delay the finding of real solutions to the problem.

Simply put, wealth destroyed by chasing non-solutions is no longer available to invest in real solutions.

Before discussing specific proposed solutions that I have placed in the "dumb" category, we should first briefly address three overriding principles that seem to guide the developers of dumb solutions to global warming.

Dumb Assumption #1: We can reduce emissions and not hurt the economy

This mistaken notion arises when people ignore the basic economic truths I summarized in Chapter 6. Proponents of the various schemes to punish the use of energy believe that the resulting pressures on industry to come up with new energy technologies will actually stimulate economic growth. While such policies will, no doubt, lead to new job opportunities for a few people, it will be at the expense of everyone else.

If such a technique works, why not spur economic activity in the construction industry by destroying all of our cities and towns? Why not spur growth in the medical community by purposely giving everyone a disease?

You cannot punish something that people do as a necessary part of daily commerce, and then expect the economy to benefit from it. Anytime economically useful activity is thwarted or punished, the economy as a whole suffers. For every Nobel Prize laureate in economics that advocates any such position, there are ten more who know that it is foolishness.

Dumb Assumption #2: The Precautionary Principle

The Precautionary Principle (PP) is often lifted up as guiding philosophy that should be followed when society considers developing some new technology. The PP states that new technologies that might risk doing damage to human health or the environment should be avoided. While this sounds like a laudable goal, it has one practical problem: no one lives their life that way.

Humanity has elevated itself above the misery, discomfort, disease, and premature death suffered by previous generations through the development of new technologies, all of which have inherent risks. Even the decisions we make as part of our daily lives, no matter how trivial or mundane (e.g., crossing the street, eating), involve balancing benefits against risks.

That the PP is so popular is simply a reflection of how risk-adverse our modern culture has become. Both the risks and benefits of new energy technologies need to be considered when deciding which new technologies to develop and utilize. We want our electricity to be continuously available, but don't want any more power plants to be built. We want more wind power, but not if that means having towers cluttering our view. We want our trash and garbage taken away, but don't want any more landfills. We like the nonpolluting aspect of nuclear power, but we don't want to deal with the waste disposal problem and security risks.

The people who advocate the PP are the ones who either want us to give up modern life, or at the very least wring their hands and complain that there has to be a better way. They do not recognize the comfortable—indeed wealthy—position from which they voice that opinion. In their imaginary world, we can have what we want, without risks, if only we try hard enough.

For some reason, this seems to be the attitude of many university professors and intellectuals. When one spends so much time dreaming about theoretical solutions to the world's problems, it is easy to confuse fantasy with reality. The practical use for such views goes no further than their bumper stickers: "Imagine World Peace." "Imagine No Pollution." How about, "Imagine Reality"?

People who actively campaign to raise awareness of the global warming problem tend to believe in the PP. But notice that, since they are so risk-adverse, they offer no realistic alternatives to fossil fuels—unless you really do yearn to live in a cave. (Again, a problem in economics, since the demand for caves would outstrip the supply.) These folks complain about the problem, but they are clueless when it comes to realistic solutions to the problem.

The PP is a guiding philosophy that unrealistically assumes we can have benefits with no risks. Since it is so widely assumed to be true by the environmentally enlightened and university educated, you need to at least be aware of its existence. But as a realistic philosophy to guide our decisions regarding possible global warming solutions, the PP is doo-doo. Flush it and forget it.

Dumb Assumption #3: Global Warming Hurts the Poor

Drawing upon the basic economic concepts I covered in Chapter 6, let's review some of the rampant misconceptions about the effect of the environmental policies on the poor. Environmentalists claim that the policies they advocate will end up helping the poor of the world—those who are the most sensitive to climate change. For instance, they claim that warmer weather in the tropics could cause the spread of some diseases.

It is vitally important that we understand that by far the greatest (and unnecessary) risk to the world's poor is *poverty*. Climate will change, with or without the help of humanity, and the best way to insulate the world's poor from natural dangers in the environment is to help them lift themselves out of poverty. Wealth generation requires access to affordable energy. If we destroy wealth in our meager attempts to prevent some theoretical future warming, we will actually be shooting ourselves in the proverbial foot.

If the poor live in a coastal area that is threatened by hurricanes, the answer isn't to pass global warming legislation that might reduce average hurricane wind speeds by 3 mph over the next hundred years. The answer is to help the poor lift themselves out of poverty to the point where they can simply escape in a reliable car when the threat arrives.

If we really want to help the world's poor, we need to be encouraging activities that actually use more energy, not less. This is what will prevent the most death and suffering in poor countries. And for those who disagree with me, I suggest they renounce their use of electricity to demonstrate their commitment to their position.

One example of the environmentalists' claim that global

warming will hurt the poor is through the spread of malaria. When it comes to malaria, though, the environmentalists should be the last ones we turn to for advice. The world's poor are already dying *by the millions* because of misguided environmental policies. European countries have threatened trade restrictions on African countries if they use DDT, a relatively safe and extremely effective pesticide that the developed countries have already used to conquer malaria. As a result of this ban, nearly one million Africans die each year from malaria. Many more are permanently disabled. Forcing the environmental policies of wealthy countries on the poor countries has caused, and continues to cause, death and suffering.

I am not claiming that there are no health or environmental risks associated with the use of DDT. I am saying that the use of a small amount of this very effective pesticide has benefits that far outweigh its dangers. If environmentalists really are interested in helping the world's poor, let them demonstrate it by publicly supporting current efforts to reinstitute the residual spraying of homes with DDT in Africa. It is indefensible that the website of the Environmental Defense Fund actually brags about their role in banning the use of DDT.

After completing one of my congressional testimonies, an African man approached me and asked who I knew to help him bring legal action against organizations whose environmental policies have led to the deaths of countless poor people in his country. I didn't know what to tell him. This situation is nothing less than a crime against humanity; it verges on genocide. I can guarantee you that the wealthy countries of the western world would not put up with an outbreak of malaria. Fortunately, after many years, a few African countries are now once again instituting residual spraying of DDT in homes, with dramatic reductions in the number of cases of malaria.

For some strange reason, westerners seem to believe that the world's poor are better off living in poverty. On one occasion, I was talking with an African at the Ugandan Embassy about how the industrialized countries view poor African countries. He told me that Africans are tired of being denied access to the global economy simply so white people can travel around Africa in their air-conditioned Range Rovers and observe the villagers' quaint and simple lifestyle. While we might yearn for the simplicity of the Africans' way of life as depicted in a *National Geographic* article, it is unlikely that we would choose to trade places with them.

Environmentalists have actually succeeded in halting plans to construct hydroelectric dams in Africa and India that would have provided electricity to people who don't yet have any. They were worried that the ecology of the river would be disrupted.

The lack of wealth generation in poor countries is actually a greater risk to the environment than is manmade pollution. Poor countries that have only wood and animal dung available as fuel to heat and cook with end up denuding the land. For instance, from satellite photographs you can vividly see the international boundary between Haiti and the Dominican Republic. Haiti is extremely poor, and most of the trees have been cut down for fuel. As a result of burning all of this wood and dung in homes and huts, over one million of the world's poor die each year from respiratory illnesses. Ah, the simple life ... sounds kind of romantic, doesn't it?

Contrary to popular opinion, the poverty in poor countries is not from a lack of natural resources. If that was the case, Japan wouldn't be a major world economic power, and Russia would be the most prosperous nation in the world. Instead, the biggest impediment to wealth building within a nation is governmental interference and control over people's lives.

This is why democratic reforms are so desperately needed in so much of the world. Political and economic freedom should be basic human rights that are universally recognized and defended. Only in this way can poverty be greatly reduced and the human condition improved. But as long as the U.N. continues to favor tyrants and dictators as model leaders for the rest of the world, I doubt that they will be helping to make this goal a reality.

The most important thing we can do for the world's poor is to support the spread of freedom, which then allows them to lift themselves up out of poverty. Environmentalists instead want to impose policies that do just the opposite—perpetuate poverty.

The proposed solutions to global warming that punish energy use will hurt the poor first, in both poor and wealthy countries. While the middle and upper classes would consider higher fuel prices to be an annoyance, higher prices can be economically devastating for the poor. Those who live from paycheck to paycheck cannot afford a doubling of the price of gasoline, heating oil, or electricity.

Partly because of the elitist, misguided views of a minority of radical environmentalists, the world's poor are denied the benefits of wealth that few, if any, of those same environmentalists would be willing to part with. You won't find these environmentalists among the one to two billion people on the Earth who still do not have access to electricity, nor among the three billion people who struggle daily to get enough food, fuel, and clean water just to survive.

No, you will find these environmentalists living in comfortably heated and cooled homes, with clean water, food, refrigeration, and good medical care. These are all comforts that are the direct result of mankind's smart and efficient use of natural resources, and from which environmentalists can have the luxury of worrying about environmental issues.

Why are we so quick to accept the claims of environmentalists about the supposed environmental effects of modern life on the world's poor? Maybe it is because of the widespread angst and guilt that so many prosperous people experience when they are reminded of the billions of people who do not have what they have. We assume that our negative effect on the environment is just one more reason to feel bad about our prosperity.

This view is not only unnecessary, it is dangerous. As I addressed in Chapter 6, wealth building is the only way to protect humanity from the daily threats posed by Mother Nature. This was true before global warming became the *cause du jour*, and it will continue to be true. The benefits of wealth and modern life far outweigh the costs, and the world's poor are eager to participate in an economic process that we seem to have taken for granted.

The United Nations

There is no better place to start than the United Nations when it comes to ineffective approaches to eliminate global warming; there is no other organization that is more efficient at spending money to not solve problems.

The U.N. has formulated and orchestrated a wide range of environmental policies over the years. United Nations bureaucrats were emboldened by the success of their 1973 Montreal Protocol. That treaty was developed to phase out the use of chlorofluorocarbon refrigerants, which are believed to be responsible for stratospheric ozone depletion. Much less harmful alternative refrigerants, as well as new designs for refrigeration equipment, were developed to take their place. For prosperous countries, the extra cost of the necessary equipment to accommodate the new refrigerant was absorbed fairly painlessly. For poor countries that were just reaching the point of being able to afford refrigeration so they could reduce the death rate from food-borne illnesses, tough luck. The price just went up.

Now the U.N. has set its sights on global warming. Global warming is, after all, the ultimate international problem, and so it is only natural that the U.N. would want to spearhead the international response to it. Unfortunately, the U.N. seems to have a history of proposing policies that put the U.N. in control of matters and which involve large transfers of wealth from richer nations to poorer nations.

Literally as I write this, there is a news report from the United Nations University that predicts that desertification and environmental degradation will result in fifty million "environmental refugees" by 2010. Well, deserts and other natural habitats have always been changing—expanding or shrinking—and mankind's best way to deal with those changes, or any other natural hazards, is to either protect ourselves from the environment or to move out of the way.

How do other people survive who choose to live in deserts or frigid climates? They have sufficient wealth to deal with it. They don't just sit outside, at the mercy of the elements, and hope that sufficient food, water, and shelter will magically drop in their laps. We have the technology for humans to live under just about any environmental conditions that nature can throw at us, and most of it is not rocket science. But this technology requires that people be free to engage in economic activity between themselves and with the rest of the world.

Why can't the United Nations spearhead efforts to spread the political and economic freedoms that are so desperately needed by the world's poor? When you see pictures of emaciated children in a dry desert location in Africa, you can bet it is not because of a lack of food. It is because of governmental policies or deliberate acts of warfare that has prevented food from reaching the people. In today's world, famines are almost never the result of a lack of food. Sure, a drought might make a food supply situation worse, but the solution is to remove the political and economic barriers that prevent food from reaching these people in the first place.

Even though Americans send billions of dollars in aid to Africa, much of that money is siphoned off by corrupt governments to help them remain in power. The rock musician Bono's desire to help poor Africans is admirable, but we have already learned that throwing more money at their problems can do more harm than good.

It appears that the United Nations is not interested in fixing, or even understanding, the source of such problems. They push environmental policies that destroy wealth rather than create wealth, and as a result prevent people from lifting themselves out of poverty. Those same policies just happen also to increase governmental control over people's lives.

I suspect I really do know the reason why the U.N. is not interested in solutions to humanity's problems. If the people of the world are empowered to solve their own problems, the U.N. bureaucrats will no longer have a job. It is as simple as that.

The Kyoto Protocol

Building upon the bureaucratic success of the Montreal Protocol for reducing the manufacture of ozone-depleting chemicals, now the United Nations is spearheading efforts to reduce mankind's emissions of greenhouse gases. The 1995 Kyoto Protocol (which is easier to say than its official name, the United Nations Framework Convention on Climate Change) is the current global warming treaty in force, ratified by all industrialized countries except Australia and the United States. The treaty is supposed to reduce global greenhouse gas emissions by an average of 20 percent below 1990 levels by 2012.

The treaty was negotiated in concert with periodic scientific summaries of our state of knowledge about manmade global warming. This scientific review body is called the Intergovernmental Panel on Climate Change (IPCC), and it produces an updated report every few years. You sometimes hear the IPCC referred to by the media as "over 2,000 climate scientists that all agree that global warming is a serious problem." In truth, most of those 2,000 "scientists" are actually bureaucrats and governmental representatives; very few of them are climate scientists.

And no one actually polled any of the scientists to ask them to agree to any such statement on global warming. Instead, a handful of politically savvy scientist-bureaucrats use the IPCC as a scientific cover to promote policies for which the science just happens to be the latest justification.

While the body of the IPCC scientists' technical report is actually pretty thorough and even-handed on the global warming issue, I doubt that any policymaker has read it all the way though. Even I haven't read it all the way through. Instead, there is a short "Summary for Policymakers" that is meant to serve that function. A handful of policy wonks who already decided what they wanted the executive summary to say produced that portion of the report.

There are even stories of these bureaucrats trying to make sure the scientific report didn't say anything that would contradict the policymaker's summary, which was written even before the scientists were done with their part.

The Summary for Policymakers is artfully worded in such a way that it does not obviously contradict the science, yet it manages to convey the maximum amount of alarm over global warming. You might remember that James Hansen's 1988 congressional testimony for Al Gore started this particularly effective approach. Any uncertainties associated with predicting climate change are either downplayed or ignored. Potential natural sources of climate variability are treated only superficially, and the bulk of the report deals with a variety of estimates of what will happen to the climate system for various assumed future scenarios of manmade greenhouse gas emissions.

As previously mentioned, after the Kyoto Protocol was devised by the United Nations, the U.S. Senate in 1997 had a brief attack of rational thought and action. They unanimously passed a resolution, 95-0, stating that they would not ratify such a treaty because of both the huge negative economic impact it would have on the U.S., and the fact that the treaty would exclude the large and most rapidly developing economies of India and China.

You might disagree with that Senate resolution, but you sure can't ignore that level of bipartisan unity. The level of congressional mass sanity exhibited by that vote is not likely to ever occur again on the subject of global warming. A truly historic event.

The Kyoto Treaty's lengthy negotiation process has provided a grand opportunity for government bureaucrats to travel to exotic locations around the world and dine in luxury. From my experience with the ones I have met, these folks believe it is their mission in life to tell you how to live yours. While you labor to support them, they love to negotiate with your freedoms. They would never consider the possibility that *what* they are negotiating might be a bad idea. Remember, more governmental regulation is good, less is bad. Just keep repeating that.

Not surprisingly, the fact that so many nations of the world have signed on to the treaty is due not so much to an altruistic desire to save the environment, but more to political and financial motives. The United Kingdom has already made a large shift toward less-polluting natural gas, and would be able to meet reduction targets relative to 1990 emissions levels more easily than would many other countries. Russia reluctantly agreed to the treaty late in the game only after the European Union agreed to allow Russia membership in the World Trade Organization in exchange.

Most poor countries saw the treaty as a wealth transfer mechanism where the wealthy industrialized countries would buy the right to pollute from poor countries through emissions trading schemes. They, personally, would not have to abide by any reduction targets—just wait for the money to start rolling in. For some reason that old phrase "follow the money" comes to mind.

The treaty finally went into force in early 2005. But by late 2005 it was already becoming apparent that most of the industrialized countries that had signed on to the treaty would not meet their emissions reduction targets. Not by a long shot. The negative impacts on business are being increasingly felt, power outages are starting to occur in Europe, and the bureaucrats are learning a lesson in basic economics the hard way.

And even if these countries do meet their emissions reduction targets, it has long been understood by everyone that the resulting effect on global temperatures would likely be unmeasurable. The Kyoto Protocol was simply a small baby step in the direction of the huge leaps that would be needed to really make a difference —say a 50 percent reduction in future greenhouse gas emissions.

The greatest frustration I have with the Kyoto Treaty process is the amount of time, energy, and wealth that was wasted on ideas that were contrary to basic economics. Everyone supports reducing our production of greenhouse gases—until they are told how much it will cost the economy. Maybe that's why they are never told.

U.S. POLICY SOLUTIONS

Congress has had a nasty habit of passing legislation based upon either lobbying groups who have co-opted some like-minded experts, or the testimony of an actor who happened to play an expert in a movie. As my wife likes to point out, scientists are usually a little understaffed in the common sense department, and so we can get mixed up in all kinds of policy battles.

There is actually a support group for such experts—the Union of Concerned Scientists. "Hi, my name is Jason, and I'm a concerned scientist." "Hi, Jason!"

Al Gore has proposed a variety of policy approaches to the global warming problem. It is too early to tell whether any of them make enough sense to contribute substantially to solving the problem without doing even greater harm from unintended consequences. But if Gore's suggestions at the end of his movie *An Inconvenient Truth* are any indication, we are simply in for more feel-good gestures. Turning off the lights when you leave the room, using compact fluorescent bulbs, and buying a hybrid car might make sense economically, but don't be fooled into believing that they are going to make any measurable difference in future global temperatures.

While these feel-good conservation approaches seem to be popular, I suspect that it is only because we like to feel good.

With the Kyoto Protocol doomed to failure to either meet its goals for emissions reductions, or to reduce future warming even if it met its goals, a couple of our senators have had the bright idea to go ahead and push for a similar plan for failure here in the United States.

The McCain-Lieberman Climate Stewardship Act, also called "Kyoto-lite," would be a wonderful opportunity to make it look like the U.S. is doing something about global warming without actually having to accomplish anything toward that goal. As of this writing (2007) it has failed to pass, but its supporters still hold out hope. If it does pass, there will be a lot of self-congratulatory pats on the back. But, like the Kyoto Protocol, there will be essentially no warming-reducing gain to show for the resulting economic pain.

The state of California, always a trendsetter for the rest of America, in September of 2006 passed the Global Warming Solutions Act. The goal of this legislation is to cap greenhouse gas emissions at 1990 levels by the year 2020. Apparently not content with the current number of businesses fleeing the state to escape overly restrictive regulations, Governor Schwarzenegger spearheaded this unprecedented legislation that environmentalists have hailed, and against which business has railed.

The Act has been called a "job killer" by the California Manufacturers and Technology Association, California Farm Bureau Federation, and California Chamber of Commerce. Just as businesses that experience overly restrictive federal policies will move their operations to other countries, businesses that experience undue state-mandated regulatory burdens can simply move to other states. For those of us not in California, but who are interested in the economic growth of our own states, we applaud California's bold, new legislation. Bring it on.

Who knows? Maybe the ultimate intent of California politicians is to drive all producers of carbon dioxide out of the state. California will mostly be a mecca for either tourism (leave your car at the Nevada border, please), or for making more movies about how evil corporations are trying to destroy the environment. It will probably go unnoticed that Californians still need the goods and services produced by those industries that had fled the state. At least their state can't be blamed for the pollution, right?

Or, as I mentioned previously, it could be that Governor Schwarzenegger plans on resurrecting his Terminator character one more time, and travel back to the 1940s to fix this problem before it even starts. Maybe what we are seeing in California is just the beginnings of an elaborately orchestrated movie plot. I think I'll wait for the DVD this time.

Besides mandating reductions in energy use, what other effective ways are there to fail at fixing the global warming problem? Let's take a look at a few silly ones.

SILLY SOLUTIONS

Sometimes I get letters from people who have a new, innovative idea to produce abundant amounts of renewable energy. For instance, putting solar collectors on our cars. Even though we would only get enough energy to reach peak speeds of 3 mph at noon on a sunny day in June, and could only drive far enough to visit the neighbor (the one who lives down the hill), at least we would be Saving the Earth.

I recently received a letter from an elderly inventor promoting a wonderful new idea to generate electricity from the wind. He even has a brochure. But he won't reveal anything about this new technology. He is looking for an investor or manufacturer, and he's afraid of someone stealing his idea. I suppose if I had such a revolutionary invention, I'd be afraid of the idea being stolen, too. I wish him success. Oh, wait ... I wonder if he knows it's already being done?

To be fair, I will admit that we need to be supportive of research into all energy technologies that have some hope of providing economical alternatives to petroleum and coal. In fact, I have even included hydrogen, solar, and wind in the next chapter (Less Dumb Global Warming Solutions). My only point here is that economically competitive alternatives are much more difficult to develop than one would think. As *Sesame Street*'s Bert once told Ernie, "It's easy to have ideas. It's not so easy to make them work."

Our fear of hurricanes, especially after the very active 2004 and 2005 hurricane seasons, has led to some discussions on how to control these out-of-control monsters. For all our technological advancement, we still can't tame a hurricane—although we have tried. Back in the 1960s, Project Stormfury researchers tried seeding the clouds in hurricanes out away from the eyewall in an attempt to form an outer, weaker eyewall. Cloud seeding, for instance with silver iodide, helps to turn super-cooled cloud water into ice. Ice particles are, in turn, necessary to form much of the precipitation that occurs in hurricanes. Unfortunately, it was discovered that virtually all of the water above the freezing level in hurricanes was already frozen, so the effort was futile.

Ideas such as towing a giant iceberg from Antarctica to the Gulf of Mexico to help cool it have been floated (ha! floated!), but it has been calculated that the iceberg would mostly melt during transit. Besides, the Gulf of Mexico is simply too large for even a giant iceberg to have any appreciable effect anyway.

Another idea is to cover the water with a liquid that would prevent evaporation. You might recall that water vapor is the fuel for hurricanes, and the warmer the water, the more evaporation there is. But since evaporation is the primary way in which a body of water gets rid of excess heat, we can imagine what would happen. Instead of naturally cooling through evaporation as Nature intended, the ocean water would continue to warm. Eventually, we would have 100° Fahrenheit sea surface temperatures, the liquid we had put on the surface would have dissipated, and then we'd get our first Category 6 hurricane. It's not nice to fool Mother Nature.

Someone even suggested detonating environmentally friendly nuclear weapons (environmentally friendly?) in strategic locations within the hurricane. But a mature hurricane is already releasing the energy equivalent of one Hiroshima-class nuclear weapon every second, so that would probably be fruitless as well.

With the power of a hurricane coming your way, I think the best idea is simply to go someplace else for a couple of days, especially if, like New Orleans, you are located below sea level. We will just have to face the fact that destructive hurricanes have always existed and always will. There isn't much we can do except prepare for their arrival.

A few proposals for creating an Earth-orbiting sun shade have been discussed. Unfortunately, putting one of any substantial size in orbit around the Earth would be hugely expensive (ha! hugely!). In case you haven't noticed, the Earth is a pretty big place.

Another sun shade idea, one that might actually have some merit, is to dump sulfate aerosols into the stratosphere like explosive volcanic eruptions do. The 1991 eruption of Mount

Pinatubo in the Philippines is estimated to have injected millions of tons of sulfur dioxide into the stratosphere, reducing the sunlight absorbed by the Earth by two to four percent for a couple of years. The sulfur dioxide gets converted to sulfuric acid aerosols, and these reflect some of the sunlight back to outer space. This method of cooling the Earth, of course, would be blocked by environmentalists, who would howl at the idea of mankind purposely polluting the atmosphere. That, after all, is the job of volcanoes.

Sustainability

The environmental movement has adopted a new buzzword, "sustainability," as justification for reducing our consumption of natural resources. Sustainability pervades environmental discussion these days and is increasingly becoming part of governmental planning and regulations. The basic idea of sustainability is that our present rate of use of most natural resources is not sustainable, and that we should work toward making it sustainable.

On the face of it, sustainability sounds like a legitimate goal. But to the extent that some of our natural resources are truly limited (coal, petroleum), it is obvious that their continued use cannot be sustained forever. If a resource is finite, then we will run out eventually, no matter how little we use at a time. The only way to avoid totally depleting it is to stop using it completely. All we can do for nonrenewable resources is to make them last a little longer.

So sustainability raises some important questions. Is there any point in making a limited resource last, say, 10 percent longer? Future generations will undoubtedly have developed new technologies that greatly reduce our dependence on these resources. Fortunately (or unfortunately, if you are worried about global warming), for every twenty years that we use petroleum, we find more than twenty years of additional supply. In fact, so many new oil finds are occurring that there are increasing numbers of geologists that don't believe all of it could have come from ancient forms of life. There is simply too much of it.

And many resources, for instance many metals, will never be

used up. The quantities in the Earth's crust are simply too abundant. Even if the massive amounts in the ground start getting scarce, at some point it will be easier to go retrieve them from where we eventually put many of them after we were done using them: in landfills.

To the extent that sustainability is a useful thing for humans to practice, free markets provide the best mechanism for that to happen. As a specific resource is gradually depleted, its cost goes up. This leads to the development of new technologies that then become more cost competitive. For instance, overfishing of ocean waters has led to fish farming in manmade ponds. Then, as a result of reduced pressure, this allows ocean stocks to replenish themselves gradually.

The adaptation to dwindling resources that free markets automatically provide then prevents the resource from ever getting used up. That so many people fret over dwindling natural resources is one more example of the simple linear thinking that leads people to believe that global warming will be a serious problem. They see a current trend, and extrapolate it far into the future, not realizing that there are other forces at work that help to stabilize the system.

Sometimes I wonder whether the simplistic, linear thinking that causes some people to worry about the climate system is the same simplistic thinking that also prevents them from understanding the self-regulating nature of the free market economic system. It is tempting to accuse such people of being simple-minded. Hmmm.

Similar types of linear thinking lead to the oft-reported claim that species are going extinct every day due to the human pressures on ecosystems. This scientific theory is definitely not a scientific observation, and is little more than an urban legend. New species are still being discovered by biologists every year. How is it that we could know there are no remaining members of a species anywhere in the world, but then we discover a new species we never knew existed? No wonder so many people are confused about science.

While mankind is fully capable of doing damage to natural habitats, we would probably be unable to drive any species to extinction even if we tried. But if we ever do try, my vote would be to start with mosquitoes.

CONSERVATION

Sustainability's little brother, conservation, actually makes some economic sense. As a potential global warming solution, though, it is relatively ineffective. About all conservation can accomplish is to slow the rate of increase in atmospheric carbon dioxide by a tiny bit.

In a free market economy, people already conserve at a level that has been balanced with the rest of their costs of living. People don't wantonly waste resources, but they also do not spend inordinate amounts of time devising ways to reduce their energy consumption. They have other things to do with their time than to figure out how to minimize the number of miles they drive each day.

In a free market economy, there is an inherent profit motive for finding cost effective ways to conserve energy. That is why "energy intensity" in the United States (the amount of energy consumed per unit of gross domestic product) continues to fall every year. But simply slowing the rate at which we produce carbon dioxide through conservation will not solve the global warming problem.

While I am completely supportive of new technologies that help improve energy efficiency, we must make sure that these technologies are achieved at reasonable cost (or, preferably, by even saving money). While gasoline-electric hybrid vehicles have helped to improve energy efficiency, they are still comparatively expensive. Nevertheless, their cost can be expected to decline gradually with time. Some hybrid cars boast very high gas mileage ratings, but the hybrid technology by itself, which recaptures energy in the form of electricity that is usually lost during braking, saves only about 15 percent on gas mileage. Additional mileage gains simply arise from having a smaller engine.

To be sure, the hybrid owner can claim to be helping to reduce carbon dioxide emissions. But for most of these vehicles that boast high gas mileage, it is more the result of a smaller engine than it is the hybrid technology itself. Even if every car in the world ends up being a hybrid, global carbon dioxide emissions will continue to rise.

Compact fluorescent light bulbs represent a technology that makes economic sense. Although they are more expensive than incandescent bulbs, they produce the same amount of light for less than one-third the electricity. They last a lot longer, too, especially in my house where upstairs foot traffic keeps blowing out light bulbs attached to the ceiling of the first floor.

These technological advances should be applauded, if only because they improve efficiency and lower costs. But we should not be fooled into believing that they will solve the global warming problem. In order to do that, we need to find abundant sources of energy that do not produce carbon dioxide.

Batteries not included.

Chapter 9: Less Dumb Global Warming Solutions

UNLESS YOU ARE committed to the idea that mankind should start living in caves and teepees, the solution to the global warming problem (to the extent a problem exists) ultimately lies in a single realm: technological advancement.

The economist Julian Simon has called mankind's creative and technological genius to solve problems our "ultimate resource." As mankind pushes back the frontiers of science and technology, we find better and cheaper ways to make the products that people need and want. And that creativity is the only hope that mankind has for substantially reducing carbon dioxide emissions.

So what do we need to do to make this happen? As it turns out, we already are doing much of what needs to be done. The United States government invests billions of your tax dollars in all kinds of energy research. Private industry likely invests even more than that.

These research and development efforts, however, are only made by countries that have generated sufficient wealth to afford them. Those countries created their wealth by allowing their citizens the freedom to benefit economically from their new ideas and hard work. In contrast, most of the currently proposed policies for "doing something" about global warming destroy wealth and are ineffective. It is counterproductive to impose policies that offer only economic pain for little warming-reducing gain.

One might wonder, what happens if, as a result of widespread economic growth around the world, we suddenly experience shortages of energy. What if economic growth explodes in the developing nations, and global energy demand rises above our ability to produce it? While admittedly painful in the short term, this demand will actually help to bring those new technologies online even sooner.

As long as demand exceeds supply, high energy prices will hasten the development and use of new energy technologies. Competing energy sources that were too expensive before the shortage then become more cost competitive. This is the basic reason why mankind will never run out of petroleum, or any other natural resource, for that matter. At some point, it simply becomes too expensive to extract from the ground compared to less expensive alternatives.

The increased profits that energy companies enjoy during energy shortages enables the private sector to pursue those newer energy technologies more aggressively. Remember, these companies want to make money, and if there is profit potential in solar, wind, hydrogen, clean coal, or new technologies that we cannot even imagine at this point, they want a piece of that action.

Following are brief summaries of the major energy technologies, in no particular order, that appear to have some potential for substantially reducing carbon dioxide emissions. The treatment is not meant to be complete or exhaustive. And if history is any lesson, some new and unexpected technology will emerge in the coming decades that will greatly reduce our dependence on all other forms of energy.

NUCLEAR POWER

It is generally accepted that a combination of the 1979 release of the Jane Fonda film *The China Syndrome* and the coincidental Three Mile Island nuclear power plant accident only a few days later led to America's current aversion to nuclear power. This double-whammy on public perception has had a devastating effect on America's nuclear power industry.

But now it is time to objectively re-evaluate the role of nuclear energy in the American energy mix. It is unlikely that a dramatically new energy technology will come online in the next twenty to thirty years, but a dedicated effort to reinstitute nuclear power in the United States could make a sizeable dent in our use of fossil fuels. It cannot happen quickly, since it takes about ten years to license and construct a nuclear power plant. But new, safer, and less expensive reactor designs have been developed in recent years which will help reduce many of the previous risks and costs. It is somewhat ironic that many of the same progressive, environmentally conscious people who point to France as one of the countries we should model our country after would shudder at adopting France's method for generating electricity. Currently, about 75 percent of France's electricity comes from nuclear power plants; in the U.S., it's 19 percent

Unfortunately, public perceptions of safety problems and the resulting regulatory requirements on plant construction have put nuclear power on our energy black list. While any push to expand our use of nuclear energy would meet with opposition, even some environmental organizations are now admitting that the risk would be lower with nuclear power than the risk of global warming from fossil fuel use. Furthermore, if hydrogen fuel cell vehicles ever become widely used, a source of energy will be required to produce the hydrogen fuel, further shifting our energy production toward the electricity sector.

Clean Coal

Coal-fired power plants currently produce over one half of America's electricity. Coal reserves are abundant in the U.S., and if coal could be burned more cleanly, then the potential threat of global warming from this source of carbon dioxide would be reduced.

Sequestering (capturing and storing) carbon dioxide is one technology that is being researched. There are a couple of experimental power plants that are sequestering the CO_2 during coal burning. The CO_2 can be stored underground, for instance pumped into caves or petroleum deposits.

This is an evolving technology that is still relatively expensive, but such problems will likely be solved with time. It is estimated that the first, near-zero emissions coal-fired power plants will come online by about 2025.

Hydrogen Power

So, what is taking so long to achieve this "hydrogen economy" we keep hearing about? There is a hydrogen mine in the next county over from me, just waiting to be used for something. I think. Well, maybe not. Oh, now I remember, it takes *energy* to create hydrogen from water. Where will that energy come from? Maybe we could use the hydrogen-powered fuel cells to generate the electricity to make more hydrogen! We could call it a "perpetual motion machine."

One of the biggest obstacles to the widespread use of hydrogen as fuel is the fact that there is no naturally occurring, readily available source of hydrogen. Oh, sure, there is plenty of hydrogen all around us—but it just happens to be tied up right now, doing other things. While water contains abundant amounts of hydrogen, it is tightly bound to oxygen (H_2O), and so it takes energy to separate it from water.

If hydrogen power ends up being widely used as a way to

avoid emitting carbon dioxide, it will likely require much greater amounts of electricity, which probably means nuclear and clean coal (see above). It has been estimated that for hydrogen to fuel our transportation needs, the electrical generating capacity in the United States will need to approximately double.

Also, there are still some technological and practical challenges with hydrogen-powered cars. For instance, hydrogen is very flammable and presents greater dangers in the event of an accident than does gasoline. Gasoline will burn only over a very narrow range of air-to-fuel mixtures. Hydrogen is flammable over a much broader range. Thus, there will be greater dangers during refueling of a hydrogen-powered vehicle compared to a gasoline-fueled vehicle.

Also, the energy content of hydrogen is relatively low. In order to store an amount of hydrogen in an automobile that will provide the traveling distance available with gasoline, it must be compressed to very high pressures, further increasing risks during refueling and collisions.

Presumably, the technological challenges and risks will eventually be reduced to the point where hydrogen-powered transportation makes sense. In general, though, this technology is not as ready as many people think it is.

Solar and Wind Energy

The allure of getting energy directly from these clean sources has always been strong. After all, the sun is what has energized all of our other sources of energy. While the use of solar power is steadily growing, it still remains a very small part of our energy mix, less than 1 percent, and it will remain so for a very long time to come.

Despite its small role, however, I have listed it in the Less Dumb Solutions chapter because it does make sense in certain applications, and because there is still the possibility that new advances (example described below) will make it more cost competitive, and thus more widely used.

The biggest problem with solar electric power remains its low energy density. The amount of sunlight that needs to be gathered to produce a substantial amount of electricity requires large areas of solar collectors. Nevertheless, there are inherent advantages of solar power. There are few, or no, moving parts for photovoltaic (PV) collectors. They can be built, and thus distributed on any scale—even for individual homes. PV technology is still relatively inefficient (about 15 percent) at converting the sunlight into electricity, but research continues into improving that efficiency.

In contrast, thermal solar systems, which can provide much of a home's domestic hot water needs in relatively sunny climates, have much higher efficiencies—closer to 90 percent. And, like solar systems, they can be implemented on a home-by-home basis. From an economic standpoint, this provides the consumer with control over the decision whether to use this source of energy or not.

Two relatively new solar technologies look intriguing to me. First, solar towers (also called solar chimneys) can capture the daily production of warm air under several square kilometers of glass-covered desert land. Since warm air that is surrounded by cooler air wants to rise, this produces a wind under the glass canopy that flows toward a central towering chimney. Turbines at the base of the tower convert the self-contained wind field energy into electricity as the warm, buoyant air rises in the tower.

The total amount of energy that can be produced in a solar tower increases with the tower height. Current tower designs run about 1 kilometer high. If such a tower could be built, it would be the tallest manmade structure on Earth. As such, it would also make a very cool tourist attraction. As of this writing, Australia is the only country that is actively involved in the planned design and construction of such a facility. A 50-kilowatt concept demonstration facility was successfully operated in Spain for several years.

Second, in the photovoltaic realm, Pyron Solar has developed an inexpensive technology to focus the sun's rays on photovoltaic cells at a very high intensity, equivalent to 400 suns. At this intensity, the cells are much more efficient at converting sunlight into electricity—about 35 percent efficiency rather than the normal 15 percent. As a result of these features, most of the expensive PV cells in such a collector can be eliminated, and the few that are used produce over twice as much electricity as those in normal applications.

Like solar power, wind power is being increasingly exploited, but still on relatively small scales. The current technology is now cost competitive in some parts of the country, but like solar, large amounts of land must be covered by windmills to generate substantial energy. Most people do not want windmills near where they live since they are considered to be somewhat of an eyesore.

While government subsidies and tax breaks can help jump-start these technologies, private industry is typically reluctant to invest much in technologies that are not fairly close to being cost effective already. In cases of artificial government support, when the subsidies and tax breaks end, the technologies once again languish. We have been down this road before during the "energy crisis" of the 1970s, and the laws of economics have not changed since then. Until fossil fuels become scarcer, and thus more expensive, either the price of these alternative technologies must be brought down, or their efficiency at generating electricity must be increased.

As the manufacturing costs of solar and wind power technologies continue to fall, and if the cost of other traditional sources of energy rise, solar and wind energy use will continue to grow. But the inherent limitation of the amount of wind energy or solar energy available over the solar collector area, or the wind turbine's blade area, means that they will continue to be minor contributors to the total energy needs of humanity. Nevertheless, I have included them in the Less Dumb Solutions chapter because, if the public had sufficient will, these technologies could be deployed relatively soon on as large a scale as we would be willing to pay for—financially and aesthetically.

Biofuels

There is increasing interest in using plant matter to replace some of the gasoline and diesel fuel that we currently extract from petroleum. Because of the sheer volume of fuel we use, however, it has been estimated that all of the corn grown in the U.S. would replace only about 12 percent of our gasoline needs.

As more ethanol for gasoline (and vegetable oil for diesel fuel) is produced, there is the unintended consequence of rising food prices. If the supply of crops does not increase in proportion to demand, prices must rise. As a result, some foods become more expensive and, once again, the poor are the first to suffer.

In summary, there are a few existing alternatives to fossil fuels than can somewhat reduce our greenhouse gas emissions. But to really make a major contribution to the problem we will need major technological advances. The good news is that both the government and private industry are investing in new energy research. It will take time for these new technologies to come online, however. Nuclear and clean coal both have promise, but substantial expansion of their use will take ten years or more in the case of nuclear, twenty years in the case of clean coal.

There are no quick fixes. The smart solutions to the global warming problem will take time. It is therefore important that we do not become impatient, because it is impatience that leads to governmental policies that have a history of doing more harm than good. Again I will emphasize, when politicians start pushing for legislation to attack global warming, we must ask them two questions: "How much will it cost?" and, even more importantly, "How much will it help?"

Unless we are smart about our policies, at best we will merely have "feel good" measures that do little more than make a few politicians look noble. At worst, the time and wealth that is wasted on expensive and ineffective policies will delay the development of the technological advances that represent our only hope for greatly reducing carbon dioxide emissions.

THE FUTURE

The solutions to current environmental problems in general, and the global warming problem in particular, largely depend upon an informed public. Economically, you vote for specific goods and services with your dollars, and so you have some control over what kinds of pollution you are producing as a result. Politically, you vote for representatives who are, for the most part, going to follow the desires of their constituents in formulating public policy.

Wielding this economic and political power responsibly requires knowledge. And that, I'm afraid, is where we have a major problem to overcome.

The environmentalist agenda tends to be anti-progress, ignoring the only real solution to the global warming problem: human ingenuity and technological advancement. Environmentalists tend to appeal to our emotions when they push for certain policies, and this is dangerous because it can lead to bad decisions. Energy is a physically produced, economically driven commodity, and there is no way to avoid the realities of physics and economics when developing smart energy policies. It is time to "Imagine Reality." It is time to shout down the environmental extremists who perpetuate exaggerated views of risk and never mention benefits when discussing energy use. No one lives their lives avoiding all risk and ignoring benefits, including the environmentalists.

It is unfortunate that media reporting in the U.S. tends to be biased toward social and political agendas that perpetuate environmental, social, and economic myths and half-truths. As a result of the media's narrow views, the public continues to be misled on important environmental matters. Too many people remain unaware of the real costs and human consequences of some the currently proposed environmental policies. While a few have spoken out on the widespread public misconceptions about environmental risks (e.g., Bjorn Lomborg and John Stossel), for the most part the problem still exists.

I'm sure that journalists have good intentions, but they are apparently unaware that their ideologically biased reporting on such important policy matters can do more harm than good. It is easy to whip up public hysteria. It is not so easy to look beyond one's own biases to understand global warming science and policy issues. Fortunately, alternative media sources such as the internet and cable news are enabling more diverse views on environmental issues to be advanced and discussed. Facts and reasoned debate are necessary if we are to avoid letting our emotional attachment to some supposed solution get in the way of finding real solutions.

When it comes to environmental issues in general, and global warming in particular, the future of humanity lies in Julian Simon's "ultimate resource": human creativity. This creativity needs to be fostered and rewarded, not stifled. While working to improve the human condition, people need to be viewed more as producers and stewards, rather than consumers and polluters.

Everyone seems to appreciate the desirability of the United States becoming more energy independent. Our dependence on energy sources from politically unstable countries is very risky, and represents a strategic vulnerability. But as long as the supposed "rights" of nature supersede the rights of the people to use the natural resources that they require to thrive, the United States will never approach energy independence. You can't simply wish or legislate new energy sources into existence.

The world has immense coal reserves, possibly enough for another 1,000 years or more. If the by-products of coal combustion (e.g., mercury, carbon dioxide) that are not yet scrubbed out can be greatly reduced or captured, then mankind will have a relatively clean source of energy for a long time to come. And these clean electrical generation technologies will be needed if we ever achieve the hydrogen economy. Hydrogen will need to be extracted from water, and energy will be required for this conversion.

I predict that when these technologies are ready, environmental fears in the headlines will continue. Environmental worriers have worldviews and jobs at stake. Anything that is good for human progress is going to be portrayed as bad for the environment, period. In the real world, risk can never be eliminated, and so the worriers will never be satisfied.

The technological advances that we need will be considerably delayed if we do not encourage the continued generation of wealth. It is the wealthy countries of the world that can afford the large investments in research and technology that will bring about these advancements. Punitive policies such as mandated reductions in carbon dioxide emissions will have little impact on future global temperatures, and it could easily result in a global economic recession. This, in turn, could delay by many years the necessary technological advances we need. This is especially true in the private sector where, in the face of an economic downturn, the first place that companies cut investment is research and development.

The good news that you never hear about is that the United States government is already investing billions of your tax dollars in new energy technologies. Private industry is no doubt investing heavily as well. All of humanity requires access to affordable energy, and the need will never go away. As long as billions of the Earth's inhabitants continue to try to elevate themselves above poverty, there will be a continuing growth in energy use. And as long as there is a desire for cleaner energy, new technologies will provide our only way of getting there.

Chapter 10: Summary

WE ARE RAPIDLY entering an age where too much free time, too much faith in the ability of science to predict the future, and too little spiritual fulfillment are leading too many people to believe in pseudo-scientific predictions of environmental disasters. As the mother of all these threats, global warming is now perceived to be the ultimate global crisis against which all mankind must unite. There is a religious fervor that accompanies this belief, and as a result we are now scaring ourselves (and our children) to death with the new state-supported religion and its teachings of mankind's sins against Mother Earth.

I have nothing against people's religious beliefs—only their labeling them as "science."

Environmentalists, politicians, movie stars, and the media all want you to believe that currently proposed solutions to global warming will save the Earth, help the poor, and keep humanity from destroying itself. This book has explained why I believe that the Earth's climate system is not nearly as sensitive to humanity's greenhouse gas emissions as many scientists think it is.

But even if those scientists are correct, and dangerous levels of global warming await us, the solution to the global warming problem will not be found in the Kyoto treaty, or in any of the policy changes currently being proposed in congress. Our only real hope of substantially reducing greenhouse gas emissions will be through our "ultimate resource": human ingenuity.

THE SCIENCE

Belief in catastrophic global warming has little scientific basis, and perpetuates the bad habit that scientists have of predicting environmental doom. Great significance is attached to some short-term change that is observed by science, for instance a change in the amount of ice in the Greenland ice sheet or increased melting of Arctic sea ice in the summer, and then it is extrapolated far into the future. Long ago, in 1874, Mark Twain noted this bad habit of scientists when he wrote,

> There is something fascinating about science. One gets such wholesale returns of conjecture out of such a trifling investment of fact.

The science of climate change is still in its infancy, and most climate scientists still do not appreciate the full complexity of the climate system. While computerized climate models do indeed contain enough physical processes to mimic many average aspects of the Earth's climate, there are good reasons to believe that they do not yet contain all of the important stabilizing processes that really exist in nature. As a result, those models tend to produce too much climate change in response to the small, 1 percent enhancement of the Earth's natural greenhouse effect that will result from humanity's doubling of the atmospheric carbon dioxide concentration sometime late in this century.

In this book I have tried to explain, in as simple terms as possible, the "big picture" of how the climate system operates, and let you decide whether projections of catastrophic climate change can be believed.

Let's review the big picture. In response to solar heating, weather (wind, evaporation, precipitation, clouds, etc.) act to cool the Earth's globally averaged surface temperature to well below what the natural greenhouse effect tries to make it: 57° Fahrenheit, rather than 140° Fahrenheit. Published back in the 1960s, this is one of the first research findings regarding the operation of the climate system. As a result of this cooling, the oft-repeated claim that the Earth's "greenhouse effect makes the Earth habitably warm" is less true, quantitatively, than the fact that "weather keeps the Earth habitably cool."

Note that, at this point, we already see the bias that exists in scientists' explanation of the "greenhouse effect" to the public.

Yes, the Earth's natural greenhouse effect does make the surface of the Earth warmer than if there was no greenhouse effect. But it is not some benign, static, self-existent quantity. Dominated by water vapor and clouds, the natural greenhouse effect is constantly being adjusted by weather processes, which directly or indirectly control how much of each of those is produced.

Take Earth's dominant greenhouse gas, water vapor, as an example. Despite the continuous evaporation of water from the Earth's surface in response to solar heating, the atmosphere never fills up with it. Theoretically, nature could allow it to keep accumulating, causing a runaway greenhouse effect that would warm the Earth much more than it in fact does. Why doesn't this happen? Because that vapor is continuously kept in check by the only atmospheric process that depletes it: precipitation.

Precipitation processes act as nature's thermostat, adjusting how much vapor will be allowed to remain in the atmosphere, thereby controlling most of the Earth's greenhouse effect.

And guess which atmospheric process we understand the least? Precipitation!

Take clouds, the second largest component of the Earth's greenhouse effect, and the component that cools the Earth by reflecting sunlight back to outer space, as the second example. Many climate scientists don't realize it, but even clouds far away from any precipitation activity are controlled by precipitation processes. The millions of square miles of low stratus clouds over the cooler parts of the subtropical oceans form underneath a lower atmospheric temperature inversion (warm air layer). That inversion is caused by air being forced to sink in response to warm air rising in precipitation systems, possibly thousands of miles away. In general, it is precipitation (how much is formed, and at what altitude) that controls the vertical temperature structure of the atmosphere, and that temperature structure, in turn, influences cloud formation.

So we see that, ultimately, precipitation systems exert the largest single controlling influence on Earth's average climate. I believe that control is a thermostatic one. If the Earth gets too warm, precipitation processes change in such a way to cool it down. If the Earth gets too cool, those systems change their behavior to warm it up again.

The thermostatic control system in your house might be small and somewhat complex, but you know it must be understood in order to explain the temperature of the air in your house. Very few climate researchers are actively trying to understand how the Earth's thermostatic control system operates. It is so complex, and so little is understood about it, we just sweep it under the rug and hope that it's not too important.

As a result, the climate modelers' belief in a sensitive climate system is due to a misplaced faith in overly simplistic climate models. It takes a higher level of understanding to include in those models all of the stabilizing processes that exist in the real climate system. That climate models still have a tendency to drift away from the real climate state is evidence of this overly simplistic behavior. The models are precariously balanced on a knife-edge, overly sensitive to any disturbance such as mankind's tiny enhancement of the greenhouse effect.

In climate parlance, the models still do not contain all of the negative feedbacks that exist in nature. Like a weight hanging from the end of a spring, or a marble rolling around in the bottom of a bowl, these negative feedbacks are restoring forces which keep the system from departing too far from its average state.

Global warming pessimists will no doubt claim that I have too much faith in the existence of stabilizing processes in the climate system, processes that have not yet been discovered. I would counter that those scientists have too much faith in crude climate models. Extraordinary claims require extraordinary evidence. Even the modelers admit that clouds are still a big wild card in projections of future climate change.

I predict that there will be an increasing number of scientific publications in the coming years describing "newly found" stabilizing processes in the climate system. These stabilizing mechanisms, of course, have always existed—it is merely the scientists' discovery of them that will be new. I further predict that the most important stabilizing processes will be traced to the behavior precipitation systems.

The Political Players

If we are looking for a disastrous positive feedback mechanism in global warming, we need look no further than the interactions between worried scientists, the eager media, and pandering politicians, all of whom have vested interests.

Scientists are funded by these government programs to research the problem of manmade global warming, and so everything they find ends up being put in that context. They are supposed to be investigating the *problem* of global warming, not the lack of a problem. The vast majority of published scientific research simply *assumes* that current warming is manmade, and not substantially the result of natural climate variability. To demonstrate otherwise, we would need to identify and understand natural climate variability—which, for the most part, we don't.

Most scientists researching global warming (including me) receive a continuous flow of funding from government programs, and have built careers and theories that they would like to continue. They want to believe that their jobs are important to humanity, and that their research really will help humanity keep from destroying itself. Their emotions color their judgment when talking to reporters. Uncertainties are minimized, and sound bites are carefully constructed to sound as dramatic as possible while remaining factually correct.

The media are not unbiased, either, as their sensationalizing of the problem helps their careers. They are out to make the world a better place, and how better to accomplish that than to warn readers and viewers that it is time for us to change our evil environment-destroying ways?—in between ads for new SUVs.

Fears of catastrophic global warming and claims that the global warming science has been "settled" have been so amplified by the media that global warming skeptics like me are being increasingly demonized. I fear it is only a matter of time before congressional hearings are held to investigate why some skeptical scientists have not jumped onto the global warming bandwagon—inquisitions to pressure all scientists into having politically correct views on the subject.

Even the environmental lobbying groups are not unbiased, because their jobs are totally dependent upon the existence of threats to the environment, and what could be a bigger threat than global environmental collapse from catastrophic warming? Many of them depend upon donations from charitable foundations that do not have to answer to any public desires or priorities, just the whims and political biases of the foundation board members. And guess what? Many of them also get money from (gasp) Big Oil!

Although they claim to hold the moral high ground on the issue, professional environmentalists have more to lose than anyone if the global warming problem goes away.

Politicians recognize that their power and influence can be enhanced by getting involved in the global warming debate. There will be huge winners and losers financially as a result of any legislation to curb carbon dioxide emissions. Unfortunately, some of these politicians are simply pandering to widespread public opinion that we need to "do something now," despite the fact that we are already spending billions of dollars on new energy research and technologies.

On the international level, the United Nations' desire to control the affairs of nations has never been closer to reality, now that the world is supposedly faced with a global environmental catastrophe. Most of the countries of the world that have signed the Kyoto Treaty have only done so in anticipation of large transfers of wealth to them from the developed countries. The wealthy countries will, in effect, be paying the poor countries for the right to pollute.

Even some major corporations are planning on what they consider to be inevitable governmental controls on carbon dioxide emissions, and they want to position themselves to fare better than their competitors. Thus, British Petroleum (BP) becomes "Beyond Petroleum." Follow the money.

I don't want to make it sound like everyone involved in the global warming debate has corrupt motives. I merely want to dispel the myth that any one of the players can claim the moral high ground. Everyone is biased by their own self-interests. The widespread practice of demonizing global warming skeptics simply because some (but not all) of them might have received some limited funding from private industry is hypocritical, and is little more than an *ad hominem* tactic that allows the demonizers to avoid having to discuss the real issues. In reality, the real money has been made by several high-profile global warming alarmists who have received large, no-strings-attached monetary awards from left-leaning philanthropic foundations. No such right-leaning awards exist.

The Policies

Our inability to deal with global warming policy in a realistic manner is partly due to our modern risk-adverse culture. This risk adversity is, in the end, more dangerous for humanity than the risk itself. It is time to start standing up for the benefits of modern technology and energy use when confronting those who would only complain about the risks. Those who complain only do so from the health, comfort, and convenience of their modern lifestyles.

As John Stossel has pointed out, when we give a disproportionate amount of our finite wealth to some exaggerated threat, we can literally end up "scaring ourselves to death." Media hype over the global warming issue might sell magazines and increase viewership, but it has the power to kill people. Anytime we divert wealth to misguided policies because of public sentiment based upon misinformation, that wealth is no longer available to address more important problems.

And, remember, it is only the vibrant and growing economies of the world that can afford the research and development activities that will lead to cleaner sources of energy. Only the wealthy countries of the world can afford to clean up their environmental messes. Unfortunately, carbon taxes and international income redistribution schemes like the Kyoto Protocol instead destroy existing wealth and prevent the creation of new wealth. These punitive policies then become economically counterproductive, possibly even delaying the development of the new energy technologies that we need.

The most infamous example of the unintended negative consequences of environmental policies based upon exaggerated fears is DDT, a relative safe and very effective pesticide used to stop the spread of malaria by mosquitoes. Knee-jerk reactionary bans on DDT by many countries are directly responsible for up to one million malaria deaths in Africa each year. As long as this modern-day holocaust is ignored by the mainstream media, I will continue to accuse them of being more concerned with the radical environmentalist agenda than they are with alleviating human suffering. They certainly do *not* hold the moral high ground.

Some will ask, "But shouldn't we greatly reduce our production of greenhouse gases—just in case? After all, we buy insurance to protect the investment we have in our homes." Sure, if it was that easy, that cheap, and if we had any assurance that the insurance policy would actually pay up if we ever had to make a claim. Unfortunately, most of the currently proposed "solutions" to the global warming problem are both expensive and ineffective, and so the analogy to insurance for those solutions is a poor one.

Substantially reducing humanity's emissions of carbon dioxide will be extremely difficult for at least the next twenty years. One of the most basic human needs is access to affordable energy, which then allows people to engage in a variety of activities that are necessary for humanity to thrive. Like it or not, human emissions of CO_2 will continue to rise during this time no matter what we do. Billions of people in the world are just now lifting themselves up out of poverty, and they will not stop just because a scientist's computer program says they should.

The economic policies that most of the global warming alarmists advocate are now failing to achieve their goals of emissions reductions. The European Union is learning that it is not so easy to simply mandate the reduction of carbon emissions based upon the desires of bureaucrats. Within one year of implementation, the Kyoto Protocol was mostly failing at ever reaching its goal of reduced emissions by 2012. Furthermore, the environmentalists' pressure against the construction of new power plants will very likely lead to energy shortages there in the coming years.

And even if the emissions reduction goals were obtained, the effort is so feeble that the effect on future global temperatures would be unmeasurable anyway. And now, some in the United States Congress seem intent on to emulating this failure with a variety of bills that are even weaker than the Kyoto treaty.

Some states such as California are not going to wait for federal legislation; they are claiming to be taking the lead on emissions reductions. But if they are successful, they will merely force polluting industries to move elsewhere.

It is not sufficient that environmental goals might have been born of noble intentions. While good intentions to help both humanity and the environment are laudable, we must be smart about our policy actions. People need to understand that the only benefit these policies will have is to make ourselves feel better that we "doing something" about the problem. What matters are results, and we have had a long enough history of making mistakes to enable us to start making more intelligent decisions.

THE PHILOSOPHICAL MOTIVATIONS

Modern environmentalism in general is couched in buzz words and terms that sound noble, but end up being hypocritical attempts to keep humanity from prospering. I'll have more sympathy for environmentalists' calls for draconian solutions to the global warming problem when they have stopped using automobiles, airplanes, electricity, modern medicines, and all the other benefits that a strong market-based economy has brought to their lives.

Environmentalist's invocation of the "precautionary principle" is nothing more than a stealthy ploy to prevent further human development. People do not actually live their lives and make their decisions based upon this principle, because it unrealistically assumes we can have benefits with no risks.

Similarly, sustainability is an illusion that also stifles economic progress. For the few natural resources that are truly limited, the only way to avoid running out of them is to stop using them altogether. Fortunately, human ingenuity combined with free markets always finds a way to provide goods and services that require a minimum of scarce resources. This is simply because scarce resources become expensive compared to alternative resources and technologies.

In general, environmental policy decisions that favor nature over people are based upon worldviews or religious beliefs that are separate from the science. Like it or not, humans must alter their environment to fit their needs. It is what we do, and we should not apologize or feel guilty about it. Science doesn't care what we do about our environment—only people care. Any rights that we confer upon nature through environmental policies should only be those that benefit humanity. Anything else verges on a state-supported Pagan religion.

It seems that many of those who support the currently proposed global warming policies carry around some sort of self-loathing angst over their prosperous positions in the world. But if we are not smart about our policy decisions, that angst over environmental problems could be replaced with the angst that one to two billion of the poorest people in this world must endure on a daily basis. Their children are dying from disease; from spoiled food due to a lack of refrigeration; from mosquito bites because their country has been threatened with trade sanctions if they use DDT; or from respiratory disease and death due to smoke from the indoor burning of wood and dung. Entire countries are being denuded in the search for more wood. These are a few of the sources of angst for most of the world's poor.

The Good News

The good news is that, even if global warming ends up being a real problem, we are already "doing something" to solve the problem. New and cleaner ways of providing the energy that humanity needs are now being actively researched and developed. The U.S. government is investing hundreds of millions of your tax dollars each year in new energy technology research. Private industry is also investing in research, knowing that there will be great profit potential for anyone who develops new energy technologies, since everyone needs energy.

To the extent that global warming will be a problem, it is human ingenuity and the development of new energy technologies that will solve that problem. Any efforts that divert us from technological advances not only waste time and wealth, but also threaten personal health and well being. Unless you are part of that radical sect of environmentalists that wants modern society to be destroyed, new technologies are the only hope for the carbon-free energy sources we seek.

We are now approaching a decision point. Do we want to solve the global warming problem, or just pretend we are doing something about it? Do we want humanity to thrive, or to wither? As the calls for action to fight global warming become louder and more shrill, it is imperative that the public start asking two critical questions: "How much will the proposed solution cost?" and "How much future warming will it alleviate?" We must not let the pushers of harmful and impotent policies get their way with feel-good platitudes and generalities about "addressing" the problem.

I often wonder: What motivates politicians and environmentalists who advocate policies that are not only doomed to failure, but also hurt so many other people ... especially the poor?

I'll leave it up to you to figure that one out.

Epilogue

IMAGINE . . .

. . . somewhere in Africa there is a six-year-old girl playing with her best friend in a small, remote village. Tragically, she will not live to see her seventh birthday. In three months, malaria will snuff out her short life. Her country has been prevented, through economic threats made by environmentally conscious foreign countries, from using a small amount of pesticide on the doorposts of her family's hut. As a result, during the night while she sleeps, a mosquito will inject her with deadly malaria.

The rest of the world will never benefit from what this little girl had to offer humanity. Thirty years hence, as a professional chemist working for a major petroleum company, she would have spearheaded the development of a revolutionary new energy technology that humanity desperately needed. But instead, only death awaits her.

Your voice, empowered by knowledge, is what humanity now needs to keep us from continuing to sacrifice innocent lives at the altar of radical environmentalism.

Illustration Credits

Page 1 – Clive Goddard; p. 11 – Mike Williams; p. 35 – Mike Baldwin; p. 45 – Joel Mishon; p. 62 – Ralph Hagen; p. 85 – Clive Goddard; p. 103 – Nik Scott; p. 124 – Mike Baldwin; p. 139 – Fran; p. 160 – Mike Baldwin; p. 171 – Glenn Foden.

Index

ABC News online, 31
Africa, 3, 23, 89, 94–95, 99, 144, 178, 183; hydroelectric dams in, 145
Alabama, hurricanes in, 113
Antarctica, icebergs in, 155; ice sheet, 1
Arctic Ocean, 76; sea ice, 82, 172
Atlantis, 32
Atmospheric circulation system, 45–61, 84
Australia, 27, 149

Bangladesh, cyclones in, 115
Barbour, Haley, 19
Bell, Art, 134; *The Coming Global Superstorm*, 2
Bermuda, 20–23; Triangle, 32
Bernstein, Carl, 25
Bible, the, 100–102
Big business, 136–138
Big Oil, 6–7, 40, 117–119, 136, 176
Biofuels, 167
British Petroleum (BP), 177
Bush administration, 122, 129–131
Butterfly effect, 48
Byrd-Hagel Resolution, 131

California, viii, 153, 179; Chamber of Commerce, 153; Farm Bureau, 153; Manufacturers and Technology Association, 153
Carbon credits, 140
Carbon dioxide, atmospheric, 17, 74, 78, 81, 158; emissions of, 1, 62–67, 80–83, 114, 135, 159, 161, 179; and warming tendencies, 70–71, 83
Carrey, Jim, 28
Carson, Rachel, *Silent Spring*, 3
Castro, Fidel, 112–113
Cathedral of St. John the Divine (Manhattan), 101
Cess, Bob, 77
China, air pollution in, 74, 131, 150
China Syndrome, The (film), 162
Christy, John, 16
Clean Air Act, 135
Climate, and boundary value problem, 48; changes in, 74–75; definition of, 47; models, viii, 67–84, 85–93, 175; stabilizing mechanisms in, 80, 174–175; system, 45–61, 172–175; *see also* Global warming
Clinton, Bill, 134
Clouds, seeding, 155; systems, 59–60, 72–74, 79, 173–175
Coal, clean, 161, 163, 167; reserves, 169
Compact fluorescent light bulbs, 159
Condensational heating, 58–60
Conservation, economics of, 158–159

Coriolis effect, 61
Corporations, research funded by, 6–7
Crichton, Michael, 98–99
C-SPAN, 133
Cuba, 112–113
Cyclone Larry, 27

Day After Tomorrow, The (film), 2
DDT, in Africa, 3, 89, 94–95, 99, 144, 178, 181; in America, 3
Department of Energy, 7, 125
DiCaprio, Leonardo, vii
Discover (magazine), 88
Dominican Republic, 145

Earth, Day, 16; resilience of, 2, 171; thermostatic controls of, 45, 78–80, 173–174
EcoEnquirer.com, 31–33
Economics, and conservation, 158–159; and energy shortages, 161; and global warming, 8, 103–123, 141, 168; and pollution, 119–123, 151; and sustainability, 157
Ehrlich, Paul, *The Population Bomb*, 3–4, 89
11th Hour, The (film), vii
Environmental Defense Fund, 144
Environmentalism, 88–102, 143, 180–181; and economics, 116–123
EPA, 7, 120–121, 124, 135–136
Ethanol, 167
European Union, 151, 179
Evaporation, 56–57
Exxon-Mobil, 95
ExxonSecrets.org, 6

Feedback, 67–68, 75, 174
FEMA, 19
Fonda, Jane, 162
Fossil fuels, alternatives to, 117, 139–141; burning of, viii, 1, 86, 88,
France, 162
Frank, Neil, 20
Free market economies, 157–158

Gaia hypothesis, 99–101
Gates, Bill, 111–112
Germany, after World War I, 106; Green Party in, 2
Global warming, catastrophic, 25; denial of, 93–95; economics of, 8, 103–123; federal expenditure on, 1; history of, 36–37; and infrared radiation, 48–55, 60, 64–69, 77, 81; manmade, vii, 9, 40, 80–83, 175; policy responses to, viii, 9, 89, 95, 177–179; politics of, 124–138, 175–177; and the poor, 143–147; possible causes of, 62–84; as religion, vii–viii, 8, 95–102, 171; role of government in, 8, 88; solutions to, 139–170; water vapor increases and, 70–72; *see also* Climate
Global Warming Solutions Act (California), 153
Gore, Al, 2, 6, 16, 20, 94, 132–134, 150; *Earth in the Balance: Ecology and the Human Spirit*, 97, 132; *An Inconvenient Truth*, vii–viii, 11–12, 40, 152
Gray, Bill, 20
Greenhouse effect, 52–54, 70–71, 73, 77
Greenhouse gases, viii, 14, 53, 62–65, 76, 81, 92, 151, 175
Greenland, ice sheet, 1, 172; Vikings in, 36–82

Greenpeace, 98
Grist (online magazine), 94
Gulf of Mexico, 111, 118, 155
Gulf Stream, 1, 30, 69

Hagel, Senator Chuck, 134
Haiti, 145
Hansen, James, 7, 129–130, 132–133, 150
Heartland Institute, 85
Hurricanes, 12, 18–23, 154–155; Florence, 20; forecasting, 20; Katrina, 18–19, 113
Hybrid vehicles, gasoline-electric, 158–159
Hydroelectric dams, 145
Hydrogen power, 154, 161, 163–164, 169

Ice Age, Little, 15; new, 1, 13
India, air pollution in, 131, 150; hydroelectric dams in, 145
Indonesia, 2004 tsunami in, 17–18
Initial value problem, 47–48
Intergovernmental Panel on Climate Change (IPCC), 149; Summary for Policymakers, 149–150

Japan, cancer rates in, 122; natural resources in, 145
Jones, Phil, 133
Journal of Geophysical Research, 71

Kennedy, Robert F., Jr., 19
Kerry, John, 7
Kerry, Teresa Heinz, 7
Kurishio Current, 69
Kyoto Protocol, 95, 125, 149–152, 172, 177–179

Landsea, Chris, 20, 131
Little House on the Prairie (TV show), 115
Lomborg, Bjorn, *The Skeptical Environmentalist*, 4–5, 115, 168
Louisiana, hurricanes in, 18–19, 113
Lovelock, James, 99

MacArthur Foundation, 7
Maher, Bill, 131
Malaria, 3, 99, 144, 178
Margulis, Lynn, 99
Marshall, Barry, 42
Martin, Steve, ix
Mauna Loa Observatory, 63
Mayfield, Max, 20
McCain-Lieberman Climate Stewardship Act, 152
Media, environmental alarmism in, 25–31, 176
Medieval Warm Period (Medieval Optimum), 13, 15, 36
Meteorologists, 86–93
Microsoft, 111–112
Mississippi, hurricanes in, 18–19, 113
Montreal Protocol, 147
Mount Pinatubo, 43, 96, 155–156

NASA, 7, 17, 100, 125, 128–130; Goddard Institute for Space Studies, 7; Mission to Planet Earth, 128; satellites, 65–66, 128, 133; Sojourner, 134
National Academy of Sciences, 15
National Acid Precipitation Assessment Project (NAPAP), 135
National Enquirer, 27
National Geographic, 145
National Hurricane Center (NHC), 19–23
National Oceanic and Atmospheric Administration (NOAA), 7, 125
National Press Club, 134
National Religious Partnership for the Environment (NREP), 101

National Science Foundation (NSF), 125
Nature (journal), 29–30
Nature, quasi-religious views of, 5, romantic notions of, 5
Newsweek (magazine), 27
Nixon, Richard, 25
Nobel Prizes, 7, 25, 42
Non-Governmental Organizations (NGOs), 7
Nuclear power, 162, 167

Oreskes, Naomi, 44

Pacific Ocean, 76
Paganism, 97–98, 99–100, 101, 180
Paleoclimatology, 14–15, 36
Peptic ulcers, 42
Philippines, the, 43, 156
Polar bears, 40
Precautionary principle, 141–143, 180
Precipitation processes, 45–61, 173
Project Stormfury, 154–155
Pulitzer Prizes, 25

Research, government-sponsored, 7, 125–130, 160, 167, 170, 181; industry-sponsored, 160, 167
Robbins, Lionel, 104, 154
Royal Society of London, 95
Russia, and the Kyoto Protocol, 151; natural resources in, 145

Sagan, Carl, 99
Schell, Jonathan, 88
Schneider, Stephen, 88–89
Schwarzenegger, Arnold, vii, 153
Science (journal), 29–30, 77, 98
Science in, biases in, 37–40; uncertainty in, 41–44, 87–102
Scientific journals, review process of, 28–29
Sea surface temperatures, 23
Second Law of Thermodynamics, 55, 76
Sesame Street, 154
Simon, Julian, 3–4, 160, 169
60 Minutes, 94, 106–107
Smith, Adam, 109–110
Solar power, 117, 139–140, 154, 161, 164–166
Soviet Union, economics of, 110
Sowell, Thomas, 105
Spears, Britney, 28
Stossel, John, "Are We Scaring Ourselves to Death?," 107, 168, 178
Stove, David, *Darwinian Fairytales*, 64
Streisand, Barbra, 18–19
Sustainability, 156–158, 180
Sweden, Social Democrats in, 113

TechCentralStation.com, 6
Thermohaline circulation, in the deep oceans, 30
Three Mile Island, 162
Time (magazine), 26–27
Truman Show, The (film), 28
Twain, Mark, 1, 172

Uganda, 144–145
Union of Concerned Scientists, 103, 152
United Kingdom, 151
United Nations, 115, 124, 131, 145, 147–151, 176–177; Framework Convention on Climate Change, 149; University, 147
United States, area of, 16; cancer rates in, 122; climate-related research in, 27, 160, 170, 181; Congress, 131–136, 150, 179;

energy independence of, 169; energy intensity of, 158; fossil fuel use of, 162; free market economy of, 110–114, 158; and Kyoto Protocol, 149–150, 152; policy responses to global warming of, 152–153; president of, 19; record high temperatures in, 13
Urban heat island effect, 57

Warren, Robin, 42
Washington Post, 25
Watergate scandal, 25
Watson, Paul, 98
Wealth, creation of, 2
Weather, forecasting the, 20, 47–48; influences on, 5; records, 12–18; system, 45–61; *see also* Climate
White, Lynn, Jr., 98
Williams, Walter, 105
Wind power, 117, 139–140, 154, 161, 164–166
Woodward, Bob, 25
World Trade Organization, 151

A Note on the Type

Climate Confusion has been set in Nofret, a type designed by Gertrude Zapf von Hesse, the noted German Galligrapher, type designer, and book artist. Strongly reminiscent of the designer's calligraphic hand, Nofret roman is beautiful at both text and display sizes. The italic is especially spirited and elegant, and together the types contribute a lively, contemporary energy to even the simplest page of text.

DESIGN AND COMPOSITION BY CARL W. SCARBROUGH